Computer Based Projects for a Chemistry Curriculum

Authors

Thomas J. Manning

Department of Chemistry
Valdosta State University Valdosta GA 31698
United States

&

Aurora P. Gramatges

Instituto Superior de Tecnología y Ciencias Aplicadas
La Habana
Cuba

Computer Based Projects for a Chemistry Curriculum

Authors: Thomas J. Manning and Aurora P. Gramatges

eISBN: 978-1-60805-193-9

ISBN: 978-1-60805-730-6

Published by Bentham Science Publishers – Sharjah, UAE. All Rights Reserved.

BENTHAM SCIENCE Bentham 🌐 Books

DEDICATIONS

Dedicated to the memory of Bette Ann and Jim Manning, TJM's parents, for all of their love, caring and hard work. And to Uncle Bob Manning and Uncle Bob McCarthy who helped keep our extended family balanced, taught us the value of a good story, and encouraged us to push ahead. And finally to my beautiful wife Arlene and my three great kids, Morgan, Erin and Sean, for all of their love, patience and support. I would also like to thank all of the teachers that have endured me over the years at St. Joseph and St. Thomas in Pleasant Plains (Staten island, NY), Monsignor Farrell High School, College of Charleston chemistry department, University of Florida chemistry department (Dr. Jim Winefordner), Los Alamos National Lab (Bryon Palmer, Doug Hof), Florida State University (Dr. Greg Chopin) and all of the students at Valdosta State University that have pushed me to work hard and learn more to keep up with their hopes and dreams.

Dedicated to the memory of my mother, Aurora Gramatges López, who always stood by me with her unconditional love, and made me think that I could do whatever I aim to, and to Simón Rodríguez Calvo, the best teacher I can dream of. They both taught me how to be a better person and educator. And most of all, to my son Daniel, source of all my strength and inspiration.

CONTENTS

FOREWORD

Chemistry is based on experimental facts, which in this century are determined by more and more complicated techniques. In this sense, computers are unavoidable for processing experimental data. Nevertheless, the principles and methodology for experimental work remain valid. At first sight, the methodology used in the determination and use of parameters seems to be something of no interest and importance. That sin is paid at a very high price. Fortunately, books like this one save many souls. And this book has been written so well, with such a practical approach that permits students interested in Chemistry to learn easily the way to work with experimental data in order to use them fully, something extremely important in modern science.

The most important topics of modern Chemistry are analyzed by the authors from an experimental point of view. Topics that go from molecular structure, through acidity, to Nanotechnology and Radioactivity have been treated in a very fine way. Practically nothing nowadays important is missing. Direct relations between these topics and our daily life have been didactically established as a way to stimulate the interest of the readers.

I am sure that the students will enjoy reading and using this book as I did. For Teachers this book constitutes an excellent didactic instrument.

Interestingly, the authors, both excellent Chemists and Teachers, are from two countries geographically close but so "far away" for other reasons, something that could not become an unsolvable obstacle for them.

Prof. Roberto Cao, PhD
President of the Cuban Chemical Society
University of Havana
Havana
Cuba

PREFACE

In October of 2006, Thomas Manning traveled with a group of delegates from the American Chemical Society to the 27th Latin American Congress on Chemistry in Havana, Cuba, held on Oct. 16-20 at the Havana International Conference Center. (CEN, November 20, 2006, Volume 84, Number 47, p. 93, ¡Viva la Química!). During this visit he had an opportunity to meet several chemists that were faculty at universities and institutes throughout Cuba. From this scientific meeting collaboration was developed that allowed us to first participate in a joint class project involving a new exercise called Electronic Qualitative Analysis Schemes (EQAS, The Chemical Educator, 2007). The collaboration was extended first to an International Periodic Puzzle competition [1, 2] and then to this manuscript. TJM returned in 2009 to solidify the collaborations, also with a delegation from the ACS, headed by Julie Callahan, PhD the Global Network Content Development Manager for the ACS. TJM's first trip to Cuba, in 2001, allowed him to tour the beautiful island nation in a trip organized by the University System of Georgia (USG) International office. www.valdosta.edu/~tmanning/cubausg

REFERENCES

[1]   http://www.valdosta.edu/news/releases/puzzles.042710/
[2]   http://www.valdosta.edu/periodicpuzzles/results2010.php

Thomas J. Manning
Department of Chemistry
Valdosta State University Valdosta GA 31698
United States

&

Aurora P. Gramatges
Instituto Superior de Tecnología y Ciencias Aplicadas
La Habana
Cuba

ACKNOWLEDGEMENTS

We would like to especially thank a group of academic collaborators for their contributions to the development and writing of the book. They are: C. J. Mock, Caley Allen, Jeff Felton, Landon Lassiter, Peter Vu, Ryenne Ogburn, Sofia Bullah, Travis Ireland, Vineet Kumar (Valdosta State University) and Geyser Fernández Catá (Instituto Superior de Tecnología y Ciencias Aplicadas).

We would also like to thank the organizers of the 27th Latin American Chemistry conference including Dr. Alberto Nuñez, Dr. Roberto Cao and Dr. Irma Castro. The American Chemical Society is thanked for organizing the trip (Dr. Brad Miller, Dr. Beth Rudd, Dr. Jerry Bell, and Tamara Nameroff). We would like to thank Valdosta State University including the chemistry department (Dr. Jim Baxter), International Programs (Dr. Ivan Nikolov, Dave Starling), Arts and Sciences (Dr. Linda Calendrillo, Dr. Jim LaPlant, Dr. Connie Richards), Academic Affairs (Dr. Louis Levy) and Information Technology (Joe Newton, Ike Barton) that helped make various aspects of this project possible. Also, parts of this book were stimulated by grants from the National Science Foundation (NSF)-Nanotechnology in Undergraduate Education program and the Environmental Protection Agency (EPA)-P3 fund to TJM (PI) Ryenne Ogburn would like to thank the VSU-QEP project for their support.

CONFLICT OF INTEREST

The authors confirm that this ebook content have no conflict of interest.

Aim, Audience and Purpose

- This manual is a collection of spreadsheet (Excel) and Computational (Spartan) exercises.

- It is aimed at undergraduate students with a focus on general chemistry.

- This book fills a unique niche in that an entire two semester general chemistry lab course can be taught just doing the exercises presented here.

- A number of exercises could be used in a high school or AP chemistry course.

- The exercises are designed to last between three and twelve hours.

- Students can complete the exercises with minimum oversight from an instructor.

- Many academic institutions have the logistical problem of an increasing student population needing to be served with the same number of faculty, lab space, money for supplies, *etc.* By splitting a lab section into ½ wet labs and ½ computational labs, valuable resources can be conserved. Instead of 24 students doing 14 wet labs, 48 students can do seven wet labs and seven computer exercises (24 wet, 24 computer), splitting time between a we t lab and a computer lab.

- This manual recognizes that the use of computational techniques is growing in real world chemistry but many undergraduate science curriculums do not reflect this trend. It also recognizes the use of spreadsheets as a valuable tool for students simulating systems and manipulating experimental data.

- This manual can substitute for lab exercises used in distance learning courses.

Assumptions: Exercise 3 through 21 assume students have completed Exercises 1 and 2 and are familiar with basic commands of Excel and Spartan as well as with the lab write-up formats.

CHAPTER 1

Equations in Excel

Thomas J. Manning[*] and Aurora P. Gramatges

Department of Chemistry, Valdosta State University, Valdosta, Georgia, USA, and Instituto Superior de Tecnología y Ciencias Aplicadas, La Habana, Cuba

Abstract: This lab exercise aims to teach students the basic of performing calculations in a spreadsheet, the basic chemical concepts such as temperature conversion, energy units, pH and kinetics as used to learn the spreadsheet commands, the basics of generating graphs in a spreadsheet and the correct format for reports, which will be generated in a word processing program.

Keywords: Chemical concepts, spreadsheet, graphs, calculations, lab reports.

INTRODUCTION

Performing calculations in a spreadsheet is essential to analyze data from experiments in a Chemistry lab. In this exercise you will perform calculations in Excel [1], and you will learn how to generate a report in a word processing program using an adequate lab report format [2]. You will type questions and answers, generate and import graphs from your spreadsheet program, *etc.* You should pay special attention to the use of appropriate chemical terminology in the concepts studied in this exercise, such as temperature conversion, energy units, pH and kinetics [3]. The top of the first page of your report should list your *name, the date, and the lab title.* Throughout the report you should use 12 point font, number the pages in the lower right corner, and use 1 inch margins all around (Left, Right, Top, and Bottom). When entering numbers include units (*i.e.* enter 4 $^{\circ}$F not 4). Equations should be centered, numbered and the variables clearly defined in the following format:

$$P\,V = n\,R\,T$$

Where,

$$V = \text{Volume (L)}$$

---

*Address correspondence to Thomas J. Manning:** Department of Chemistry, Valdosta State University Valdosta GA 31698, USA; Tel: 229-333-7178; E-mail: tmanning@valdosta.edu

P = Pressure (atm)

n = moles (mol)

R = Gas Law constant (0.0821 L.atm/mol.K)

T = Temperature (Kelvin)

Pre-Lab Questions

Type questions 1, 2, and 3 into your report, show what equation you used to answer the question and provide an answer.

1. What is the equation for converting temperature (^{o}F) into temperature (Celsius)? Convert 72 ^{o}F into Celsius and 4 ^{o}C into ^{o}F.

2. What is the equation for converting temperature (^{o}F) into temperature (Kelvin). Convert 298 K into ^{o}F, and convert 212 ^{o}F into K.

3. What is the equation for converting temperature (Kelvin) into temperature (Celsius). Convert 25 ^{o}C into K, and 0 K into ^{o}C and ^{o}F.

Excel Graphing Exercise #1.

Temperature Conversion and 2-D graph

Save your work to at least two memory devices (*i.e.* memory stick, hard disk) on a regular basis! Many university and library computers have security programs that will delete your file if saved to the hard disk.

1. Open a new worksheet in your spreadsheet (the instructions here assume you are working with Excel).

2. In location A1 type "TEMP, oC". This is a header and is used to identify the numbers in the column.

3. In location A2 type the number "0.0". You do not type the quotation marks into your spreadsheet, just the values inside them.

4. In location A3 type the equation "=sum(a2+1)"

5. Left click in A3 so the box turns black.

6. With the black box surrounding A3, right click, hit "edit" and "copy". The box should now have a moving line around it.

7. Place the arrow in A4 and left click and continue to hold it down, dragging down to position A1000. The closer to the bottom of the screen you move the arrow the faster it will scroll down. Once you've reach your location, let the clicker up. Your numbers should go from 0-998 in increments of 1. This is much easier than entering one number at a time!

8. The A column is your temperature in degrees Celsius.

9. In the B column we will convert Celsius to Kelvin. In B1 type "TEMP, K".

10. In B2 type "=sum(a2+273.15)" This equation takes the sum in location A2 and adds 273.15 to it.

11. Left click in B2 so the box turns black, now right click and hit the command "copy". The running line should appear around the box.

12. Place the arrow in B3, hold the arrow down and drag down to B1000 and right click and paste. In the position "B1000" you should have the number 1271.15.

13. Now a simple graph will be generated using the data in A2.A1000 and B2.B1000.

14. On the top row you should have an icon labeled "Insert" click on it. Then click on "Chart".

15. Click on "standard types", "xy scatter" and sub-chart-type select the one with only dots. Then click on "Next" (near bottom).

16. Click on "Series" and click on "Remove" if any series are already present.

17. Once the series box is empty, click on "Add" and then click the arrow in the "X-values" box. The cursor should be flashing in this box.

18. Now click on the location A2, hold the clicker down, and drag down to A1000 and release. You just defined the values for the x-axis. The running rope should appear around the values.

19. Now click in the "y values" box. If any symbols are present, back space to remove them. Than block off B2 to B1000 for the y-axis.

20. Click on "Next". You should have your first view of the graph.

21. In "Chart Title" enter "Temperature Conversion. Your Name".

22. In X-axis enter "Temperature ($^{\circ}$C)". You should always enter the name of the parameter (temperature in this case) and the actual unit used ($^{\circ}$C in this case). "o" should be a superscript but if you cannot find this function in your version of Excel, leave it as "$^{\circ}$C".

23. In Y-axis enter "Temperature (K)".

24. Click on the tab "Gridlines" and remove the checks from all boxes. In some cases this is a preference unless you are trying to correlate values from the access to the trend line.

25. Click on the tab called "Legend" and remove the arrow in the Legend box. Legends are essential if you plot two or more data sets on the same graph but for a single set it is not needed.

26. Click on "Next". You can than make it a chart on the same sheet as the data (Sheet) or place it on its own Chart. Select "chart" here and click next.

27. Once the graph appears, there may be some additional manipulations to do. First, right click inside the graph but not on the trend line. Than under Area select "None". This will get rid of the gray background, saving your cartridge and giving a cleaner looking graph.

28. Place the very tip of the arrow directly on the x-axis and left click so two dots appear on either end of the axis. Immediately right click and select "Format Axis" and "Scale". Change the major unit to "100" and select "OK".

29. Now a trend line will be added that fits the straight line trend y=mx+b.

30. Click on the data points so that a number of them before highlighted. If the arrow ever becomes jammed, simply place the arrow well outside in the graph in the spreadsheet region and click once or twice than click on the trend line again.

31. Once the data points are highlighted, Right click and select "Add Trend line".

32. Under tab "type" select "linear". Than select the tab "options" and select (arrow) "Display Equation on Chart" and "Display R-squared value on chart" and hit "OK".

33. You can click on the equation and correlation coefficient and move it to an area away from the data. An R2 value of 1 means a perfect fit. You should have a slope of 1 and an intercept of 273.15.

34. Click in the box that surrounds the graph. It should become defined by black dots. Copy and paste your graph into a WORD document. Reduce its size so it is 4 inches wide by 3 inches high.

35. Type a figure caption clearly explaining the graph. This should be standard on all graphs you construct. You should now have a graph like the one in Fig. **1**.

36. Be sure to get into the habit of saving your Excel file every few minutes while working with it. Also save your WORD document.

37. In the upper right hand corner of Excel and Word is a box with text that states "Type a question for help" You should not spend all of your time learning every function that these programs have built in but you will need to learn new functions from time to time.

This is the general format of your lab report. Don't forget to include pre-lab Questions 1-3.

Joe Smith   January 15th, 2008,   Spreadsheet Exercise #1

1. What is the equation for converting temperature ($^{\circ}$F) into temperature (Celsius)?

    (Your answer)

2. What is the equation for converting temperature ($^{\circ}$F) into temperature (Kelvin)?

    (Your answer)

3. What is the equation for converting temperature (Kelvin) into temperature (Celsius)?

    (Your answer)

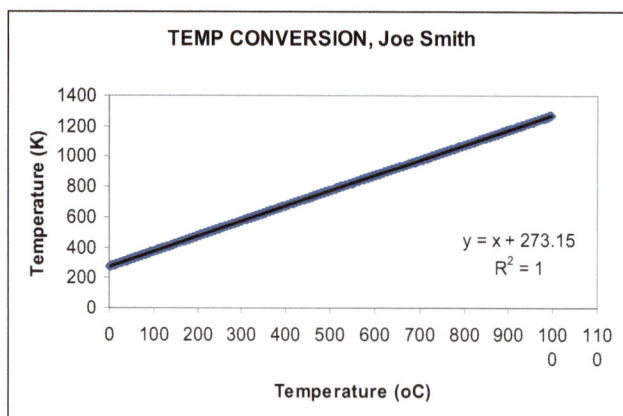

Figure 1: The correlation between the Celsius and Kelvin temperature scales. In the lower right hand corner is best fit equation (y=mx+b) and the correlation coefficient for the line.

Unit Conversions and a Three Dimensional Plot

Generate Table **1** in your report. Complete the table with the correct energy conversions. Identify the table in your report as "Table **1**: Energy Unit Conversions".

Table 1: Construct and complete this energy conversion table in your report.

1.00	Calorie (cal)
	Kilocalories (kcal)
	Watt Hours
	Watt seconds
	Kilowatt Hours (kW/hr)
	Kilowatt seconds (kW/s)
	Joules (J)
	Kilojoules (kJ)
	Ergs
	British thermal Units (BTU's)

First, you will go through the mechanics of performing a three dimensional (x, y, z axis) plot in Excel using data in the Table **2**.

a. Copy X,Y,Z values below into your spreadsheet. The titles (X,Y,Z) should start up in locations A1, B1, and C1 and the numerical values in a block defined by A2….C16. A2 defines the upper left hand boundary of the block and C16 defines the lower right hand boundary of the block.

Table 2: Data that will be used in constructing a 3D plot.

X	Y	Z
1	0	0
2	2	5
3	4	10
4	6	15
5	8	20
6	10	25
7	12	30
8	14	35
9	16	40
10	18	45
11	20	50
12	22	55
13	24	60
14	26	65
15	28	70

b.   Block off the three columns (A2…C16).

c.   Click on the "Insert" icon, click on "Chart".

d.   Under Chart type, click on "Surfaces".

e.   Click on the graphical illustration that shows a 3D plot.

f.   Click next; try clicking on both "rows' and "columns" to set how each graph appears.

g.   Under "Series" notice that it has graphed 15 series, or it has broken up your data set into 15 sets of (x, y, z) data.

h.   Click next. Under the "Titles" tab enter "Time (s)" in x-axis, "Distance (m)" in y-axis, "Chem Rules (fun)" in the z-axis, and "Your Name" as the chart title. Note – ALL GRAPHS should have your name on the top. No name, no credit!

i.   Go to the Gridlines tab and remove all of the checks, and go to Legend and remove it also.

j.   Click "Next" and "Finish". Your graph should look like Fig. **2**.

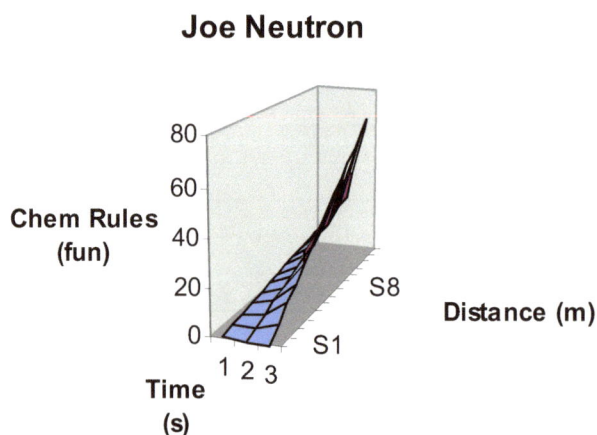

Figure 2: An example of a 3D plot.

Copy your graph into your report. It should be called "Table **2**: 3D Practice Graph". You will now create a 3D graph based on the energy units in Table **1** above. It will be called "Graph 1.3. A three dimensional graph involving energy relationships". It will be generated in your spreadsheet and copied over to your report. Be sure to include the graph label and your name on top.

The x-axis (column A) will be Joules (0,10,20,30,40,50,60,70,80,90,100). A1 will be the title slot (Joules), locations A2……A12 will be for the data.

The y-axis (column B) will be in calories. B1 will be the title slot (Calories). B2…B12 will use the equation "=sum(A2/4.184)". Enter the equation in B2 and copy it down to A12.

The z-axis (column C) will be in kiloJoules. C1 will be the title slot. C2…C12 you will convert column A to kJ using the equation "sum(a2/1000)", copy and paste it down.

Once your data is entered, be sure to label the axis, place YOUR name on top and follow the other directions above (remove legend, *etc.*). Copy and paste this graph into your WORD document. And be sure it occupies no more than 1/2 of a page. Be sure to put a brief (1 phrase) figure caption below it.

pH and [H$^+$]; Exponents and log's in Excel

The use of log, ln, and exponent functions are routine in chemistry. This exercise is aimed at introducing you to some of the basic mechanics associated with these calculations. Later in this course you will study the concepts of acidity and basicity. One equation you will frequently use is the calculation of pH from the hydronium (H_3O^+) concentration.

$$pH = -\log_{10}[H_3O^+]$$

Likewise, the [H_3O^+] can be calculated from pH using the equation:

$$[H_3O^+] = 10^{-pH}$$

pH is a unitless number while [H_3O^+] has the units of Molar (M, moles/liter) Using your calculator, fill in the following table (this table should be in your report). Identify it in your report as "Table **3**: pH and hydronium concentration scale":

Table 3: Construct and complete this table in your report.

pH	H_3O^+
1.0 (very acidic)	
7.0 (neutral)	
14.0 (very basic)	
-1 (concentrated strong acid)	
7.34 (your blood)	
8.3 (the ocean)	
4.0 (acidic, a soft drink)	
	10^{-7} M
	10^{-4} M
	10^{-10} M
	2.3×10^{-5} M
	6.78×10^{-10} M
	.00712 M

In Excel, we will enter approximately 150 pH values, in increments of 0.1, and convert them to $[H_3O^+]$ values. At the bottom of your Excel sheet, you'll see the tabs Sheet 1, Sheet 2 and Sheet 3. You can use more than one sheet.

a. Start a new sheet.

b. In box A1 enter the title "pH".

c. In box A2 enter the number "-1".

d. In box A3 enter the equation "=sum(A2+0.1)".

e. Copy this equation down to position A152. 14.0 should be the number in A152. Also note the value in A12 should be zero (0) but isn't (its -1.4×10^{-16} in mine). This is a round off error in the computer but is sufficiently small that it will never impact your calculations.

f. In B2 enter the equation "=SUM(10^-A2)". Don't forget the negative sign that comes from equation x. The " ^ " function is for raising a value to the power of ten.

g. Copy and paste the equation down to B152.

h. Block off A2.B152.

i. Start the graphing process. (select "graph" as described above).

j. Once the Chart Wizard is open (Step 1), select the "Standard Types" tab, select XY (scatter), and select the image with curved lines. Click on "Next".

k. Be sure that the "Series in" is clicked on columns.

l. Click next (you should now be in Step 3). Label the x-axis (pH) and the y-axis (Hydronium conc, M) and enter your name in the Chart Title.

m. Remove "Gridlines" and remove "Legend" and click on "Next".

n. Your graph should appear on your spreadsheet. Remove the gray background (see above).

o. Place the arrow tip directly on the x-axis and click. If you hit it properly, you should see two black dots on either end of the axis. If these appear right click on the axis until you see "Format Axis".

p. Change the scale to -1 (minimum) and 14 (maximum) and the major unit to "1".

q. If you look at your values in the B column you'll notice that most of them are extremely small (<0.001) and are not really visible on the graph. Example, you can't tell the difference between 10^{-8}, 10^{-10}, and 10^{-13} molar values.

r. Left click on the y-axis so the two dots appear on this axis and right click so "Format Axis" appears.

s.   Select the "Scale" tab. And click on "Logarithm" and click on "OK".

t.   Again, click on the y-axis so the "format axis" appears.

u.   Select the tab "Number" and then select "Scientific" and enter "2" for decimals.

v.   Look closely at the graph. Notice that much of the y-axis is negative and below the zero value. This graph looks different because students are used to looking at graphs with positive X and positive Y values.

w.   Copy and paste this graph into your word document. Be sure it's only 2.5 inches high and 3 inches wide. Give it the correct figure caption. Note, on the y-axis is the $[H_3O^+]$.

x.   In the upper right hand corner if the Excel sheet is an area to "Type of Question for Help". Type in the word "Exponent" and hit enter.

y.   Look for the command called "Power". Below the graph, briefly describe the use of this command.

First Order Kinetics and an Exponential Graph

In general chemistry you will spend time and effort studying models that describe the speed of a chemical reaction. This field is called chemical kinetics. One of the most important equations is the first order equation that allows the estimation of concentration of a reactant over time. For example, the molecule ozone (O_3) will slowly decompose to form oxygen (O_2)

$$2O_3(g) \rightarrow 3O_2(g)$$

The equation that allows you to calculate the amount of ozone left after a period of time (t) is:

$$\ln(A) = -kt + \ln(A_0)$$

where:

A_0 = starting concentration of ozone

k = rate constant (min^{-1})

A = concentration of ozone after some time (t)

ln = natural log (based on the number 2.7182818)

This equation can be rearranged to,

$$A/A_o = e^{-kt}$$

And finally to the form we will use:

$$A = A_o e^{-kt}$$

For this exercise we will assume a starting concentration of $10^{-5}M$ O_3 and a rate constant of 0.00385 min^{-1}.

1. Open a new spreadsheet. In "A1" place the title "Time (min)".

2. In A2 place the number "0".

3. In A3 type the equation "=sum(A2+1)" and copy this down to A622.

4. In B1 type "Conc A".

5. In B2 enter the equation "=EXP(-A2*0.00385)*10^-5". This is equation X above. A2 represents time, 0.00385 is the rate constant and 10^-5 is the starting concentration (defined above).

6. Copy and paste this equation all the way down the B602.

7. Use the graph command and follow the normal procedures (XY Scatter, line, remove grid lines, remove legends, *etc.* for a 2D graph). Label the x-axis "Time (min)" and the y-axis "[A]" and place your name at the top.

8. The graph you see is called an exponent decay. In this particular case the reactant (A or O_3) is decreasing in time.

9. Go to the "View" tab (top) and select "Toolbars" and select "Drawing".

10. Once this tool bar appears, select the line command. It appears as a line at an angle. Click on it. Pick the y-axis point $4*10^{-6}$ and draw a straight line from the axis to the exponential decay line (see Fig. **3**).

11. From the point on the line, make a line down to the x-axis.

12. If you click on the point, Excel will give you the (x,y) value.

13. Now pick a point on the graphed line and right click on it. A number of points on the line should become illuminated.

14. Right click and select "Add trendline". This is clearly NOT a y = mx + b or a straight line fit. It is a curve.

15. Select "Exponential", than pick the tab "Options" and "Display equation on Chart" and "Display R-squared value on chart" and click "OK".

16. Move the text to a corner of the chart away from the trend line. Your graph should look like Fig. **3**.

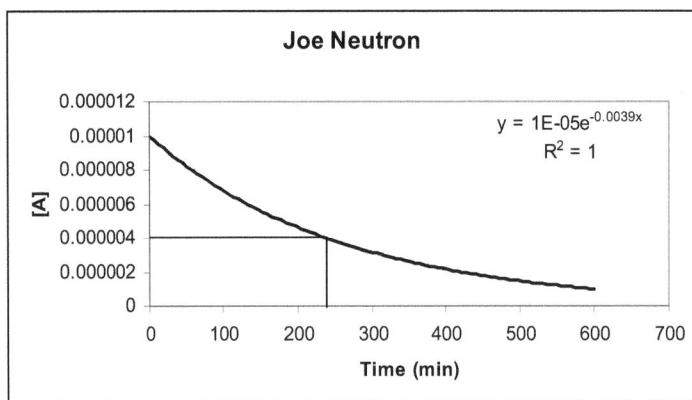

Figure 3: An exponential decay with the best fit data and correlation coefficient plotted on the graph.

These exercises should familiarize you with different aspects of performing calculations and doing graphs in Excel. Copy your graph to your report and include a properly numbered figure caption.

As you prepare your report, there are key format points to check:

1. Is your name on the top of each graph (see Fig. **3**, "Joe Neutron").

2. Are your graph axis's labeled with the description and unit (*i.e.* Time (min)).

3. Are your graphs no more than 4 inches wide by 3.5 inches high.

4. Does each graph have a numbered figure caption with a full sentence description? Are these figure captions sequentially ordered? Is the figure captions single spaced?

5. Did you answer all questions with complete sentence? Did you properly use subscripts and superscripts as needed.

6. Is your full, legal name and ID number at the top of the first page?

7. Did you insert page numbers (lower, right)?

8. Is the body of your report double spaced, 12 point font with one inch margins?

9. Does your instructor want a hard copy or a copy sent as an attachment? If a hard copy, is it stapled?

10. Does your report need references? (this one should not but some in the future may require citations).

Throughout the manual it will be assumed you completed this exercise and are familiar with the required report format.

REFERENCES

[1]   Billo, J. E. *Excel for Chemists: A Comprehensive Guide*, 3rd ed.; John Wiley & Sons, Inc., **2011**.

[2]   Rosenthal, L. C. Writing Across the Curriculum: Chemistry Lab Reports. J. Chem. Ed., **1987**, *64*, 996-999.

[3]   IUPAC Compendium of Chemical Terminology - the Gold Book, Version 2.3.1 2012-03-23. http://goldbook.iupac.org/ (accessed June 16, **2012**)

CHAPTER 2

Bonds and Lone Pairs in Small Molecules: Introduction to Spartan

Thomas J. Manning[*] and Aurora P. Gramatges

Department of Chemistry, Valdosta State University, Valdosta, Georgia, USA, and Instituto Superior de Tecnología y Ciencias Aplicadas, La Habana, Cuba

Abstract: This exercise aims to introduce students to the molecular modeling (Spartan) software. Students will construct and visualize a number of small molecules in two and three dimensions, and calculate and measure some basic geometric parameters such as bond distances and angles with three-dimensional structures.

Keywords: Chemical bonds, molecular modeling, bond distances, molecular mechanics, molecular orbital theory (MOT).

INTRODUCTION

Students will perform the lowest level of theoretical analysis (Molecular mechanics) and measure bond distances and bond angles on twenty-five common molecules [1]. Valance Shell electron Pair repulsion (VSEPR) is a model used to predict bonds, lone pairs and subsequently geometries for many small molecules. Molecular Orbital Theory (MOT) allows the prediction of bond order and the paramagnetic/diamagnetic characteristics of a molecule. A useful set of rules for the construction of small molecular species is outlined in Table **1**.

Table 1: For the construction of small nonmetallic molecules, the number of bonds and lone pairs follows the above trends (most of the time!). Following periodic trends, chlorine, bromine and iodine will often act like fluorine, and sulfur will behave like oxygen. Carbon can have four single bonds, two double bonds, a triple and a single or a double and two single bonds.

Element	Carbon	Nitrogen	Oxygen	Fluorine	Neon	Hydrogen
# of bonds	4	3	2	1	0	1
# of Lone pairs	0	1	2	3	4	0
Total	4	4	4	4	4	1

*Address correspondence to Thomas J. Manning: Department of Chemistry, Valdosta State University Valdosta GA 31698, USA; Tel: 229-333-7178; E-mail: tmanning@valdosta.edu

A bond consists of two electrons and each lone pair also consists of two electrons. Both bonds and lone pairs occupy space around the central atom. Because electrons are negatively charged, the bonds and lone pairs repel each other. Bonds are single, double or triple involving two, four or six electrons, respectively. For example, carbon in CH_4 has four single bonds, in CH_2O it has two single and one double, in CO_2 it has two double bonds and in HCN carbon has a single and a triple bond. It each case the number of bonds sums to four. Likewise, the oxygen atoms in CH_2O and CO_2 are double bonded, the hydrogen in each species is single bonded and the nitrogen in HCN is triple bonded to carbon.

Recreate Table **2** in your report, setting up a table of similar dimensions. It should have the same format and the molecules should be listed in the same order. Using the rules of thumb outlined in Table **1**, draw the structures of the molecules with the correct number of bonds and lone pairs using the appropriate 2D program. Some pointers in getting the geometry in two dimensions approximately correct are; if there are four single bonds they will be $90°$ apart; if there is a double and two single bonds they will be approximately $120°$ apart, if there are three single bonds and one lone pair they will be $90°$ apart (remember, lone pairs occupy space also!), and if you have two double bonds they will be separated by $180°$. In short, when drawing a structure if you place the bonds and lone pairs symmetrically around the central atom you will be fairly close to an appropriate two dimensional geometry (see Fig. **1**). These images can be created in a program such as "Paint", "ISIS" or WORDART.

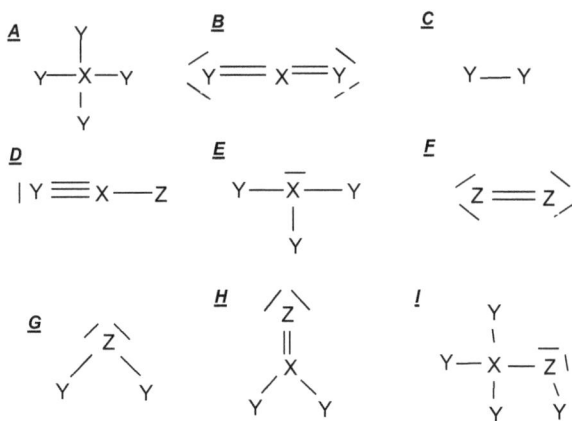

Figure 1: An example of how to draw two dimensional structures for different molecules. These structures can be related to certain molecules in Table **2**. X, Y, and Z are used to represent atoms

and the lines are single, double or triple bonds. (A) the central atom (X) has four single bonds and no lone pairs while the attached atoms (Y) have one bond and no lone pairs (B) the central atom (X) has two double bonds and no lone pairs while the two attached atoms (Y) have two bonds and two lone pairs (C) the two atoms each have one bond and no lone pairs. Two atom molecules have no bond angle. (D) the central atom (X) has a triple bond and a single bond while one of the other atoms has a triple bond and one lone pair (Y) and a single bond and no lone pairs of electrons (Z). (E) the central atom has three bonds and one lone pair (X) while the three attached atoms have one bond and no lone pairs (Y). (F) the two identical atoms (F) have two bonds and two lone pairs (G) the central atom (Z) has two single bonds and two lone pairs while the two attached atoms (Y) have a single bond and no lone pairs (H) the central atom (X) has a double bond and two single bonds and no lone pairs while one attached atom (Z) has a double bond and two lone pairs and the other two attached atoms (Y) each have a single bond and no lone pairs. (I) the central atom (X) has four single bonds and the other atom with multiple bonds (Z) has two bonds and two lone pairs. The other atoms (Y) all have single bonds and no lone pairs.

You will construct the table below in a separate document and type in (only) the empirical formulas. Call it "Table **2**. Geometries of Small Molecules in Two Dimensions". The other columns (name, structure, # bonds) will be completed with a pencil. Be sure to use subscripts on formulas (*i.e.* CH_4 not CH4).

Table 2: Construct and complete this table separately and draw the structures in a 2D program (ISIS, WordArt, paint, *etc.*) of your choice. YOUR name should be at the top of the first page of your report.

Empirical Formula	Name	Structure	# bonds # of lone pairs
1. CH_4			
2. CO_2			
3. H_2CO			
4. HCN			
5. N_2			
6. O_2			
7. H_2			
8. NH_3			
9. F_2			
10. Cl_2			
11. Br_2			
12. I_2			
13. H_2O			
14. H_2S			
15. HF			

Table 2: contd…

16. HCl			
17. CS_2			
18. CH_3OH			
19. C_2H_6			
20. C_2H_4			
21. C_2H_2			
22. CCl_4			
23. N_2H_4			
24. H_2O_2			
25. C_6H_6 (hint, it's a ring)			

There are two important points to remember when using the rules in Table **1** as a guide to build a small molecule: (1). They are rules of thumb that work with many small, nonmetallic molecules but there are some notable exceptions such as carbon monoxide (CO) and ozone (O_3). You will do molecular geometries and hybridizations in a later lab see that these rules don't always work with larger atoms or larger molecules. (2). Drawing these molecules in two dimensions do not always give an accurate image of the molecule which exists in three dimensions. You will use Spartan to construct the molecules in three dimensions and perform some simple calculations [2,3].

Be sure that your first table is completed before moving on to the next section and that it has the proper header on the first page (your name, date, experiment name).

Locate the Spartan icon on your desktop. Click on this to open the program. If it is already be open be sure to close all structures present (Click on "file" and click on "Close"). Save your work (report, Spartan files) to at least two memory devices (*i.e.* memory stick, hard disk) on a regular basis! Many computers at universities and libraries have programs installed that will delete your file saved to a hard disk automatically for security reasons. For the construction of all future structures it is important to remember to save and close your previous structure.

1. In the Upper right hand corner, click on "Options".

2. Click on "Colors". It should say "Background". You can adjust the background color. For copying and pasting these images it is best to

have a white background. Adjust the background to white now. As you build different molecules, you can use this command to adjust the color of different atoms by clicking on them (*i.e.* make all carbon atoms green, click on any carbon and use this command). Close this.

3. Now click on "File" and "New". On the right side a pad of atom choices should appear.

4. Click on the carbon with four single bonds and then click ion the middle of your work area. A carbon atom with four bonds protruding should appear.

5. Now click on the hydrogen atom with a single bond. Click on the tips or ends of the four carbon bonds, one at a time, and you should see the hydrogen atoms appear.

6. Place the arrow anywhere in your work area not on the molecule and rotate it around. This molecule (Methane) has four symmetric bonds when viewed in three dimensions. Its structure in terms of bond angles is different than the structure you drew above (Table **1**).

7. Hold the "Shift" button down and hold the right click button down and move the mouse. The size of the molecule can be made large or small using this command.

8. Under "Model" you can change the appearance of the structure. Try different appearances (*i.e.* wire, ball and wire, *etc.*). In this manual we will use ball and spoke.

9. Under "Model" click on "Labels". This will number the atoms. Depending on your color scheme/selection you may or may not be able to see these numbers. If you can't see the numbers with the white background, change the color.

10. Also under "Model" click on "Configure" and select "Mass Number". This will assign the masses (*i.e.* C=12 g/mol). This can be useful in

larger molecules with multiple elements (C, N, S, O, *etc.*) that are difficult to distinguish from each other. Once you've viewed this, remove the selection and return to the numbers assigned to different atoms.

11. Next you will minimize the molecular energy. Notice the ENERGY reading in the lower right hand corner. Click on "Build" and "Minimize". In building larger molecules you'll find the minimization command very useful in approximating the structure as you build it.

12. Click on "Setup" and "Calculations".

13. Starting at the top, select:

"Single Point energy"

"Semiempirical" and "PM3"

"Initial"

Check "Symmetry"

Total Charge "Neutral"

Compute (don't check any)

Multiplicity "Singlet"

Print (don't check any)

Click Check on Converge

Click "Submit"

Give it a name "Methane"

And click "OK" if it tells you the molecule has started and completed

14. In your report, create a table that is four columns and twenty one rows. Table **3** shows an abbreviated form of how your table will appear (its only three rows down).

15. Make your methane molecule large enough so that it fills the entire work area.

16. You can number the atoms in the display. This will be important in assigning bond angles and lengths.

17. In Spartan, click on "Edit" and click on "copy".

18. In your report, which is also open, place the arrow in the methane structure box and paste in the methane structure? Typically the structure will be too big for the box and you will have to reduce the images size to fit the box.

19. You should have Spartan and your report open at the same time.

20. Go to Spartan and click on "geometry" and "measure distance" and measure the four C-H bond distances, one at a time. Do this by clicking on the bond and recording the four distances shown in the lower right hand corner (it's in Angstroms). You'll notice in the lower right hand corner that the atoms numbers are included in the measurements.

21. Be sure to include units. If you're using WORD, you can go to "Insert" and "Symbol" and find the Angstrom symbol.

22. Go to Spartan and click on "geometry" and "measure angle". It's important to do this in the right sequence, click on a Hydrogen atom (it turns fuzzy), then click on the carbon atom and then click on another hydrogen atom. Do this for all four bond angles. If you do this in the wrong sequence (*i.e.* C, H, H) you will not get the bond angle of the central atom.

23. When you are done with your methane molecule save it, preferably to an external memory device, and close it (Click "File" and "Close"). Click on "New" so we can build a new structure.

24. We will do the same steps for CO_2 or carbon dioxide but with one twist. If you look at the element pad to the right of your work area you'll notice that there are no carbon atoms with two double bonds under the "Ent" tab. Click on the "Exp" tab and select carbon from the periodic table. Now select the linear geometry (-*-) and, beneath the periodic table, select the double bonds (=).

25. Select the "EXP" tab and select the oxygen atom with a double bond (=O). Attach the two oxygen atoms, minimize it, and rotate the molecule.

Table 3: The second table in your report will include the twenty-five molecules outlined in Table 2 - in the same order. Call this entry "Table **2**. Small Molecules in Three Dimensions; Computational Results". In molecules where the same bond produces the same angle or distance multiple time (*i.e.* see CH_4 below), only enter the respective distance or angle once.

Emp. Formula and Name	Structure	Angles, Distances
CH_4 methane		C-H(all) 1.096 Å (4 C-H bonds the same) H-C-H 109.47° (all H-C-H angles the same)
CO_2 Carbon dioxide		C=O(1) 1.096 Å C=O(2) 1.096 Å O=C=O 180°

Go through all of the steps outlined above (construct, number atoms, minimize, calculate, measure, save, close) for all twenty-five molecular species listed in Table **2**. By the time you have completed this table you should have developed mental images of some of the more common structures you will encounter in chemistry. It should also be noted that Molecular Mechanics is based on

Newtonian physics and is not considered to be the most accurate computational method. On the other hand it is much quicker than other levels of theory. This is a common trade-off in computational chemistry; typically more accurate calculations involve more powerful computers and longer computational times. Lower levels of theory can be conducted on a desktop computer in a matter of seconds. For this work the approximate results achieved with molecular mechanics are acceptable.

At the end of your report, comment on the following trends observed in your computational data. Include numbers/data to support your arguments. Name this section in your report.

"Three Dimensional Structures and Trends Observed in Calculated Bond Distances"

3a. Compare the bond lengths of the single, double and triple carbon-carbon bond in structures 19, 20, 21 using your computational results. Comment on the impact of increasing the number of bonds has on the average bond length.

3b. Compare and comment on the C-H bond distances in structures 1, 3, 4, 18, 19, 20, and 21. are they identical or different? Why?

3c. Compare the single and triple nitrogen-nitrogen bond distances in structures 5 and 23. Which is longer and shorter? Stronger and weaker?

3d. Compare the single and the double bond distances of oxygen in structures 6 and 24. Which is longer and shorter? Stronger and weaker?

3e. Compare the single and double carbon-carbon bond distances in structure #25. Which is longer and shorter? Stronger and weaker?

REFERENCES

[1]   Cramer, C. J. Essentials of computational chemistry: theories and models, 2nd ed., John Wiley & Sons, **2004**.
[2]   Hehre, W. J.; Huang, W. W. Chemistry with Computation: An Introduction to Spartan. (Wavefunction, Inc., Irvine, CA), **1995**.

[3]   Hehre, W. J. A Guide to Molecular Mechanics and Quantum Chemical Calculations (Wavefunction, Inc., Irvine, CA), http://www.csus.edu/indiv/g/ghermanb/F11_245.content/AGuidetoMM.pdf (accessed June 16, **2012**)

Send Orders of Reprints at reprints@benthamscience.net

CHAPTER 3

Titration Curve Simulating the Addition of a Strong Acid and a Strong Base

Thomas J. Manning* and Aurora P. Gramatges

Department of Chemistry, Valdosta State University, Valdosta, Georgia, USA, and Instituto Superior de Tecnología y Ciencias Aplicadas, La Habana, Cuba

Abstract: This exercise aims to reinforce in students the concepts of acid-base chemistry, specifically the reaction of a strong acid and a strong base. Students will simulate a titration curve for the addition of a strong acid (burette) into a strong base (beaker).

Keywords: Acid, base, titration, reactions, titration curve.

INTRODUCTION

In this exercise a spreadsheet is used to simulate the titration curve for the addition of a strong acid into a strong base [1,2]. Students are taken through the calculations step-by-step that mimic the neutralization reaction of hydrochloric acid (HCl) titrated into a solution of sodium hydroxide (NaOH), both in the aqueous phase. When the exercise is complete, the student will have calculated and graphed a titration curve that spans from very acidic (pH = 1) to very basic (pH = 13) regions. The two key reactions are:

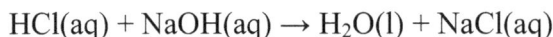

$$HCl(aq) + NaOH(aq) \rightarrow H_2O(l) + NaCl(aq)$$

$$2H_2O(l) \rightarrow H_3O^+(aq) + OH^-(aq)$$

Your lab report will include the proper header (name, title, date), pre-lab questions and answers, your titration curve (s) cut and pasted into the report, and post lab questions and answers, all typed into a single document. Your instructor may assign you an additional titration curve to calculate, graph and include in this

---

*Address correspondence to Thomas J. Manning: Department of Chemistry, Valdosta State University Valdosta GA 31698, USA; Tel: 229-333-7178; E-mail: tmanning@valdosta.edu

report (see additional exercises). For your pre-lab questions and answers, be sure to number them and show all work!

Pre-Lab questions

1. Calculate the pH, pOH, [OH⁻], [H⁺] of pure water?

2. Calculate the pH, pOH, [OH⁻], [H⁺] of 0.1 M HCl?

3. Calculate the pH, pOH, [OH⁻], [H⁺] of 0.1 M NaOH?

4. Calculate the pH, pOH, [OH⁻], [H⁺] if 50 mLs of 0.1 M HCl and 100 mLs of 0.1 M NaOH are mixed?

5. Calculate the pH, pOH, [OH⁻], [H⁺] if 100 mLs of 0.1 M HCl and 100 mLs of 0.1 M NaOH are mixed?

6. Calculate the pH, pOH, [OH⁻], [H⁺] if 150 mLs of 0.1 M HCl and 100 mLs of 0.1 M NaOH are mixed?

The name/title and pre-lab questions should take two pages (maximum) and your graph should be pasted into its own page with a figure caption (*i.e.* Fig. **1**. A graph of a strong acid, strong base titration calculated in a spreadsheet…….). You will construct the first curve following the instructions below. Once completed your instructor may provide you with a second set of titration conditions to construct your own curve. One example of this is provided after the post-lab questions. In general it will follow the same form but there may be some differences depending on what experiment you are given to simulate.

We will now go through simulating a titration curve step-by-step. The titration curve you will generate replicates the titration of a strong acid (0.1 M HCl) into a strong base (100 mLs of 0.1 M NaOH in a beaker) in one milliliter increments. A total of 198 milliliters of HCl are added so the simulated curve will properly represent regions in which both the acid and base are in excess (x-s). This exercise assumes that you have completed exercise 1, which outlines many of the basic Excel commands and outlines the details of writing a report in the correct format. Also, there are certainly ways to condense and rearrange these calculations in the spreadsheet but it is done in a step by step fashion so the student can follow each step. Remember to save your work (Excel files, report) to at least two memory devices (*i.e.* memory stick, hard disk) on a regular basis! Many computers at universities and libraries have programs installed that will delete your file automatically for security reasons.

a. In box A1 type the header "Conc. NaOH". In A2 enter the value 0.1 and copy it down to A200. This is the initial concentration of the strong base, which is in the beaker.

b. In box B1 type the header "Conc. HCl". In B2 enter the value 0.1 and copy it down to B200. This is the initial concentration of the strong acid, which is in the burette.

c. In box C1 type the header "mLs NaOH". In C2 enter the value 100 and copy it down to C200. This is how many milliliters of NaOH are in the beaker.

d. In box D1 type the header "mLs of HCl added". In D2 enter the value 0.0. Then using the command "=sum(D2+1)", which is entered in D3, copy the command down to D200. The values should increase by 1 in each box with D200 having the value of 198. This column represents the addition of 1 mL of HCl during the titration.

e. In box E1 type the header "mls of NaOH after neutralization rxn". In box E2 type the logic command " =IF(C2>D2,SUM(C2-D2)"",) ". Copy and paste this command down to E200. This should decrease by 1 in each box with one (1) being the last value observed in E101. This column represents how much NaOH has not been neutralized. After that point in the titration (E101), you have neutralized all of the NaOH and will have either a neutral solution or an excess of acid.

You are asking the spreadsheet to compare the values in C2 and D2 and determine which is larger. If C2 is greater than D2, it will perform the subtraction of "C2-D2". This is referred to as a logic command. If C2 is smaller than D2, then no value will be returned. In the upper right hand corner of Excel there is a Help command that can be used to clarify this operation or another other command we use.

f. In box F1 type the header "mls of HCl unreacted after rxn". In box F2 type the logic command " =IF(D2>C2,SUM(D2-C2)"",) ". Copy and paste this command down to E200. From F2 to F102 there should be blank spaces. In F103 you should have the number 1 and this should increase by 1 until F200 which should read "98". The numerical values in this column (1,2,3,4,5…) represent how many milliliters of 0.1 M HCl are left over after the neutralization reaction (Eq. 1,2).

g. In box G1 type "Total Volume in liters". In box G2 type the formula " =SUM(C2+D2)/1000 ". This equation adds the total number of milliliters in the beaker (C2 + D2) than divides by 1000 to convert mLs to liters. This is required because the concentrations are in Molar (moles/liter).

h. In box H1 type the header "Concentration of x-s base". In box I2 enter the logic formula " =IF(E2>0,SUM(A2*(E2/1000)/G2)"",) ". Copy this equation down to H200. This command determines if there is excess base after each addition by looking at column E. If there is excess base, than it determines the amount of base by multiply the milliliters of excess NaOH (E2) by its concentration (A2) which gives the moles of excess base (*i.e.* moles = MV). The milliliter term must be divided by 1000 in order to convert milliliters to liters. In order to convert the total moles of excess base to molarity, it is divided by the total volume in liters (G2).

i. In box I1 type the header "Concentration of x-s acid". In box I2 enter the logic formula " =IF(F2>0,SUM(B2*(F2/1000)/G2)"",) ". Copy this equation down to I200. You should start seeing real values at I103. This equation determines if there is any excess acid (column F) after the addition of the acid to the base. If excess (x-s) does exist, it calculates the molarity of the acid present by multiplying the molarity (B2) by the volume of excess (F2) to give the moles of excess acid.

The moles of excess acid is than divided by the total volume (remember you are mixing two solutions) of the solution in liters (G2).

j. In box J1 enter the header "pOH, x-s base". In J2 enter the logic formula "=IF(H2>0,-1*LOG(H2)"",) " and copy it down to J200. If the amount of hydroxide is greater than the amount of hydronium (H_3O^+) present, than it will calculate the pOH of the solution.

k. In box K1 enter the header "pH, x-s base". In K2 enter the logic statement " =IF(J2>0,SUM(14-J2)) ". This statement will convert the pOH to pH for the titration points in which there is excess base and uses the formula pH=14-pOH.

l. In box L1 type the header "pH, x-s Acid". In box L2 type the logic statement "=IF(I2>0,-1*LOG(I2)"",) ". Copy it down to L200. This command will determine if the concentration of H_3O^+ (excess HCl) is greater than OH^-. If it is, it uses the equation pH = -log(H_3O^+).

m. In box M1 type the header "Neutral". Until this point all of our calculations have deal with either excess acid or excess base. We have not yet performed a calculation in which the moles of acid and the moles of base are equal. There will be a singular addition or titration point that defines this value. In box M2 type the logic command " =IF(A2*C2=B2*D2,SUM(7*1)"",) ". Copy it down to location M200. This command will calculate the moles of acid (B2, D2) and the moles of base (A2,C2) from the equality moles = MV. If the moles are equal, than it enters a value or a pH of 7.0, which is the equivalence point.

At this point you have completed all of the calculations needed to generate a titration curve. We are now going to plot the curve in three segments on the same graph. Fig. **1** shows how your completed graph should appear (with your name on top!).

a. Click on the tab "Insert" and then select "graph".

b. Under "standard types" select "XY (scatter)", and then click on the selection which only plots points (under chart sub-type).

c. Click "Next" than select "series". Click on "Add" and then click inside the "X-axis" box. The x-axis will always be the volume added in a titration curve. Block off D2...D101 on the spreadsheet. A command should appear in the box " =Sheet1!$D$2:$D$101 ". It is often easier to block off the data you want to plot than it is to enter the command.

d. Now click the arrow in the "y-axis" box and, on the spreadsheet, block off K2...K101. In the name box enter "Excess base". Your Excel file should show the basic part of the titration curve.

e. Now we are going to plot the singular point that represents the point where the moles of acid are equal to the moles of base. Click on the tab "Add" again. Click in the "y-axis" box, then click in the box M102 (which should contain the number 7).

f. Now click in the "y-axis" box and then click on the box D102. It contains the value of acid added. In the "name" box enter "Acid=Base". This is the equivalence point.

g. Click "Add" to include one more series of values that represent the region where there is excess acid. Click on the "y-axis" box and block off from L103...L200.

h. Click on the "x-axis" box and block off D103....D200. In the name box enter "Excess Acid" and then click on "next".

i. Under Chart title enter "Titrate SA into SB, YOUR NAME".

j. Under "value X-axis" enter "Volume 0.1 M HCl added (mL)".

k. Under "value Y-axis" enter "pH".

l. Under the tab "Axes" be sure to check both Value X and Value Y axis.

m. Under the tab "Legend" be sure to check "Show legend".

n.   Click on "Next". You can now select to show the graph on the spreadsheet or a separate Chart. For this project simply select "Sheet1".

o.   You can now copy and paste this graph into your report in a Word document.

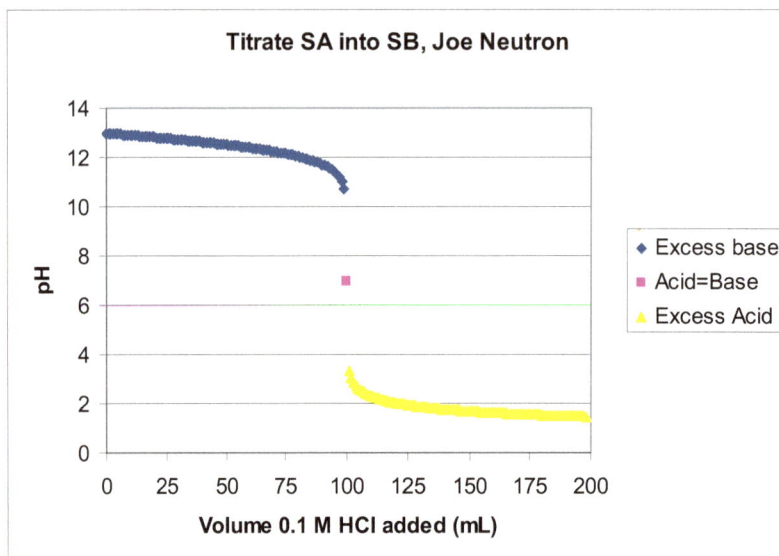

Figure 1: The completed titration curve for the titration of a strong acid into a strong base in 1 milliliter increments. It is plotted in 3 segments.

Post Lab Questions

Include these questions and answers in your report.

1.   Do strong acids and strong bases have equilibrium constants? (K_a's, K_b's). Explain. What percent of a strong acid (*i.e.* HCl) dissociates?

2.   What are the spectator ions in the titration simulated above? Do they have a significant impact on the pH after each addition?

3.   If you titrated 200 mLs of 0.1 M HNO_3 in 1 mL increments into a 100 mL solution of 0.1 M KOH, how would the resulting titration curve compare to that shown in Fig. **1**?

4. Using a computer to simulate a titration is an idealistic situation and excludes all experimental errors. Name two potential experimental errors, one operator and one instrumental, that can result in an experimental titration curve having a slightly different shape than one generated by a computer simulation.

5. Using a 2-D drawing program, sketch the shape of a titration curve in which a strong base is titrated into a strong acid (hint, the opposite of the titration you just simulated). Label the x-axis "volume" and the y-axis "pH".

Additional Exercises

1. Generate a spreadsheet for a titration curve in which 200 mLs of 0.1 M NaOH is titrated into 100 mLs of 0.1 M HCl in one milliliter increments.

2. Generate a spreadsheet for a titration curve in which 200 mLs of 0.023 M LiOH is titrated into 98.2 mLs of 0.12 M $HClO_4$ in 0.5 to 1.0 milliliter increments (you chose the increment value, but use the same value for each addition). You should have an equal number of points calculated on both sides of the equivalence point.

REFERENCES

[1]   De Levie, R. *Aqueous Acid-Base Equilibria and Titrations*, Vol. 80, Oxford University Press, **1999**.
[2]   Valcarcel, M. *Principles of analytical chemistry: a textbook*, Springer-Verlag, Berlin Heidelberg, **2000**.

Send Orders of Reprints at reprints@benthamscience.net

CHAPTER 4

Electronic Qualitative Analysis Schemes

Thomas J. Manning[*] and Aurora P. Gramatges

Department of Chemistry, Valdosta State University, Valdosta, Georgia, USA, and Instituto Superior de Tecnología y Ciencias Aplicadas, La Habana, Cuba

Abstract: This exercises aims to teach a number of chemical and physical properties for approximately 98 elements, to teach a range of periodic trends and to advance a students use of Excel and its logic commands.

Keywords: Qualitative analysis, chemical and physical properties, periodic trends, periodic table.

INTRODUCTION

Traditional qualitative analysis schemes involves the separation and identification of various water soluble ions by precipitation, odor or color changes in solution, solids or flame tests [1]. Using a simple example, separating and identifying Ag^+, NH^+_4 and Na^+ in the aqueous phase can be achieved in a three step scheme. First Ag^+ is separated and detected by adding chloride (*i.e.* KCl) resulting in a white precipitate. Second, the solutions pH is shifted by the addition of a strong base (*i.e.* KOH) resulting in $NH_4^+ + OH^- => NH_3 + H_2O$. NH_3 is more volatile than NH_4^+ and can be detected by smell. Finally the presence of the Na^+ cation can be detected with the flame test. The sodium doublet is a strong emitter of yellow light (589 nm) that can be easily observed. Because modern equipment allows for multielement analysis at parts per billion levels (*i.e.* ICP-MS) and because these schemes only deal with a small numbers of elements and emphasize very specific properties, a more rounded educational exercise is sought.

In this exercise, a group of students develop electronic qualitative analysis schemes (EQAS) for approximately 98 elements and a number of simple

---

*Address correspondence to Thomas J. Manning: Department of Chemistry, Valdosta State University Valdosta GA 31698, USA; Tel: 229-333-7178; E-mail: tmanning@valdosta.edu

molecular ions based on chemical and physical properties. Each student in the lab is assigned a group of between 4-8 species with similar characteristics. For example one student may have Li, Na, K, Rb, Cs, and Fr, while another student may have the first seven lanthanides (La, Ce, Pr, Nd, Pm, Sm, Eu). Considering there are approximately 100 elements and over twenty prominent molecular ions (*i.e.* OH^{-1}, SO_4^{-2}, *etc.*), a group of 20-22 students can cover the entire periodic table. If smaller numbers of students are involved, some chemical groups can be eliminated or the number of species provided to students can be increased (*i.e.* all lanthanides are combined into one group).

First students are given the instructions to prepare an electronic qual analysis scheme to program for the alkali metals. This will teach the concept of the electronic qualitative analysis scheme and the specifics of the programming in a step by step fashion. Once this is complete, students are given their own group of elements and instructed to construct their own electronic qual scheme. There are some basic rules to be followed and these will become more obvious as the students complete the alkali metal scheme.

The rules for developing your scheme:

1. The number of questions asked for the whole qualitative scheme should be three times the number of elements. So six elements should have a total of 18 questions. These questions will allow the participant to identify both the group they are dealing with and the specific element.

2. There should be an agreed upon reference source (or sources) that all participants have easy access to (*i.e.* textbook, Wikipedia, Los Alamos Periodic table, *etc.*).

3. Participants cannot incorporate obvious questions (*i.e.* your element has the symbol H, what is it?) or extraordinarily vague questions (*i.e.* your element has less than 150 protons).

4. There should be a minimum of three questions that allows the student to identify the group.

5. There should be at least one unique question that allows the person to identify the element.

6. There will be a minimum of one question related to electron configurations.

7. There will be a minimum of one question related to density.

8. There will be a minimum of one question on spectroscopy (light emitted or absorbed).

9. There will be a minimum of one question on physical property (conductivity, hardness, *etc.*).

10. There will be a minimum of one question on mining or mineral sources.

11. There will be a minimum of one question on related to electronegativity, ionization potential, or atomic radius.

12. There will be a minimum of one question on electrochemical properties (reduction potentials, *etc.*)

13. There will be a minimum of one question on radioactivity or isotopes.

14. There will be a minimum of one question related to solubility in a solvent.

15. There will be a minimum of one question on oxidation states in salts or water.

16. There will be a minimum of one question on industrial applications or history.

17. In rules 6-16, there may be cases where conditions are combined in a single question. For example the question, "Your element is a dication when dissolved in water, its nucleus strongly absorbs x-rays and it

will precipitate out of solution when mixed with a sulfate". BUT remember any data has to fit within the Excel box so long statements are not always practical.

18. The participant that makes up the qual scheme will also make up the numerical codes and answer keys in WORD.

19. Each element will have its own code represented by a series of 1's and 0's. The student that makes up a particular qual scheme will make up the codes for each element in their particular group.

20. Each group has its own set of codes based that appear "1011011000111110" these are entered, 1 digit at a time, in the A column (going down).

21. Be sure to adjust the width of your Excel location so all words are visible.

Using the instructions provided in this write up, enter the electronic qual scheme for the alkali group.

⇩

Review the rules for constructing an electronic qual scheme.

⇩

Your instructor will assign you or your group a set of elements to construct your own electronic qual scheme.

⇩

Research the various chemical and/or physical properties for your group and your specific elements. Use an agreed upon source or sources.

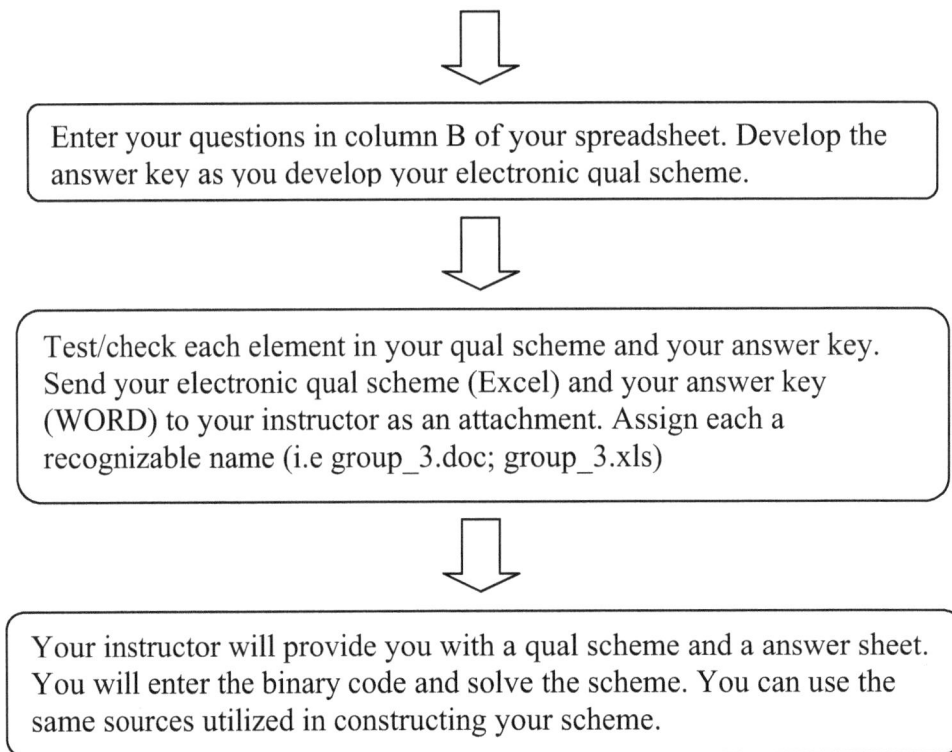

Enter your questions in column B of your spreadsheet. Develop the answer key as you develop your electronic qual scheme.

Test/check each element in your qual scheme and your answer key. Send your electronic qual scheme (Excel) and your answer key (WORD) to your instructor as an attachment. Assign each a recognizable name (i.e group_3.doc; group_3.xls)

Your instructor will provide you with a qual scheme and a answer sheet. You will enter the binary code and solve the scheme. You can use the same sources utilized in constructing your scheme.

Figure 1: A flow chart for the construction of your electronic qual scheme.

This approach has the educational advantage of covering more elements and more trends than a traditional experimental qualitative analysis scheme. It also improves a student's analytical and computer abilities. Below are the step by step instructions to construct the qualitative scheme for the alkali metals.

1. Open a new Excel Sheet. Leave A1…A21 empty. Later you will enter your code in these locations. In B1 enter the logic statement:

    "=IF(A1=1, "Your group has a +1 charge in salts"","")

    Be sure to expand the B column so the text for each answer is visible. Once the statement above is entered, you can enter the number "1" in A1 to see how it prints in B1. The above statement helps the student identify what group the element is in. All elements in this group would have a "1" as the first digit.

2. In B2 enter the statement:

    "=IF(A2=1, "Your group reacts violently with water in its neutral form","")

    This also applies to all of the alkali metals so it would be a "1" for all codes.

3. In B3 type the statement:

    "=IF(A3=1, "Your group has a +1 charge when dissolved in water","")

    Because all alkalis are M^{+1}(aq), this would be a "1" in the 3rd place in the code.

4. In B4 enter the logic command:

    " =IF(A4=1, "They are strong electrolytes when bound to the halides","")

    Since all alkali metals dissociate 100% when bound to F⁻, Cl⁻, Br⁻ or I⁻ , this would be a "1" in the 4th location.

5. In B5 type the logic statement:

    " =IF(A5=1, "Your element has a melting point of 28 oC","")

    This physical trait applies to only one element (*i.e.* Cs). So if the element is Cs, enter a "1" in this place but if it's another element enter a "0" or a "2" or another number. If this element is Cs, the code would appear as 11111. (so far), but if it's Na it might appear as 11110 or 11112 (so far).

6. In location B6 type the statement:

    " =IF(A6=1, "This element is soluble in most forms, except as a feldspar","")

Because the group has already been identified elements also found in feldspars (*i.e.* Al, Ca) would be considered but both potassium and sodium are possibilities at this point. Later questions will help narrow the choice to one. At this point, the Na or K code might appear: 111101..., Cs would appear as 111110..., and Li, Rb, and Fr would be 111100......

7. In box B7 enter the logic statement:

" =IF(A7=1, "Your element omits yellow light at 589 nm","") "

This physical trait belongs to sodium and, along with #6, helps identify the specific element. At this point Na would be 1111011 but K would be 1111010....

8. In location B8 enter the logic statement:

" =IF(A8=1, "Your element is the second least dense metal after lithium","") "

The specific value for the physical property (density) is not given forcing the student to review all of the alkali densities.

9. In location B9 type:

" =IF(A9=1, "Your element is produced by Chile and Argentina and is found in brine pools. X-6 is one its isotopes","") "

More than one element can be isolated from a brine pool (although the South American abundance helps narrow it down!) but the isotope points directly to lithium. At this point the code for Li would be 111100001....

10. In box B10 enter the logic statement:

" =IF(A10=1, "XAg4I5 has the highest room temperature conductivity of any known ionic crystal","") "

Subscripts (Ag_4I_5) are not entered in a spreadsheet header and X stands for Rb.

11. In location B11 enter the statement:

" =IF(A11=1, "Your elements outer electron is spin up - in a neutral state","") "

This statement applies to all of the elements in the alkali group because its outer electron is the s^1 (Li, $2s^1$; Na $3s^1$; K, $4s^1$; Rb, $5s^1$; Cs, $6s^1$; Fr, $7s^1$).

12. In location B12 enter the logic command:

" =IF(A12=1, "only 340 to 550 grams of your element in the earth's crust","") "

The reason why Francium is rarely mentioned in most undergraduate courses is apparent.

13. In location B13 type:

"=IF(A13=1, "Least electronegative of any known element","") "

While most academic arguments of electronegativity end with Cs, this forces students to identify the element in the lower left corner of the periodic table.

14. In box B14, enter the logic statement:

" =IF(A14=1, "Forms a compound called halite","") "

Commonly called rock salt, sodium chlorides more technical or mineral based name.

15. In box B15, enter the command:

" =IF(A15=1, "Its pure form is a grey-white metal and it readily substitutes for potassium in minerals","") "

This physical description can be applied to more than one metal but Rb does substitute for K in a number of minerals.

16. In box B16, enter the logic statement:

" =IF(A16=1, "Its chloride salt can be used to stop the heart","") "

KCl is utilized in heart surgery and lethal injections to stop the hearts rhythm.

17. In location B17 type:

" =IF(A17=1, "Its reduction potential for the M+ => M(s) is -2.925 V","") "

This forces the student to review all of the reduction potentials for elements in this group.

18. In location B18 type the logic command:

" =IF(A18=1, "Its Heat of Fusion is 63.9 kJ/mol, over ten times higher than water!","") "

A range of thermodynamic parameters can be selected (fusion, vaporization, sublimation, *etc.*).

Table 1: An example of the spreadsheet output for the element sodium. The number 1 (in column A) turns a statement on, while "0" keeps a clue hidden. The clues that are visible for a specific code allow the user to identify a group than the element.

1	Your group has a +1 charge in salts
1	Your group reacts violently with water when it's in neutral form
1	Your group has a +1 charge dissolved in water
1	Your group forms strong electrolytes with the halides
0	
1	This element is soluble in most forms, except as a feldspar
1	Your element omits yellow light at 589 nm
0	

0	
0	
1	Your elements outer electron is spin up - in the neutral state
0	
0	
1	Forms a compound called halite
0	
0	
0	
0	

Illustrated in Table **1** is the output for sodium. The code for this element would be given as: "111101100010010000". A student can use these clues to deduce that they have not only had an alkali metal but also that it is Na. While this particular flow chart focused on the group properties first and the element properties second, these questions can be presented in a completely random order. For this particular flowchart, each element would have codes similar to that shown in Table **2**.

Table 2: The codes for the different alkali elements. Each code series turns on different clues allowing the participant to deduce the element (with the proper resources).

Element	Code
Li	111100001010000000
Na	111101100010010000
K	111101010010000110
Rb	111100000110001000
Cs	111100000010000001
Fr	111100000011100000

Once this is complete your instructor will assign you a number (see Table **3**) or assign you a specific group of elements. You will develop your own electronic qualitative analysis scheme like that shown above. Be sure to use agreed upon reference sources (*e.g.* handbook [2], a General Chemistry text, Wikipedia (www.wikipedia.com), *etc.*). Review the nineteen rules and guidelines outlined in the introduction before starting your own spreadsheet. Once completed, test your algorithm and develop and answer key like that shown in Table **2**.

Once you've completed your electronic scheme they will be collected by the instructor (Scheme in Excel file and the answer key in Word). Typically they are sent as attachments. The instructor will rename and redistribute the EQS's and number codes which will serve as your answer key. You will turn these back to the instructor with your answer.

Table 3: The periodic table and a number of prominent molecular anions are arranged to form twenty two groups.

	Name	
1	Alkali metal (1A)	Li, Na, K, Rb, Cs, Fr
2	Gases (8A)	H, He, Ne, Ar, Kr, Xe, Rn
3	Alkaline earth (2A)	Be, Mg, Ca, Sr, Ba, Ra
4	Transition metals (3B, 4B)	Sc, Y, Ti, Zr, Hf, La
5	Lanthanides I	Ce, Pr, Nd, Pm, Sm, Eu, Gd
6	Metalloids	B, Si, As, Te, Ge, Sb
7	Actinides I	Ac, Th, Pa, U, Np, Pu
8	Transition metals (5B, 6B)	V,Nb,Ta,Cr,Mo,W
9	Lanthanides II	Tb,Dy,Ho,Er,Tm,Yb,Lu
10	Halogens	F, Cl, Br, I, At
11	Soft metals	Al, Ga, In, Sn
12	Nonmetals	C, P, Se, N,O,S
13	Transition metals 7B,1B	Mn,Tc,Re,Cu,Ag,Au
14	Transition metals 8B	Fe, Ru,Os,Ir,Rh,Co
15	Actinides II	Am,Cm,Bk,Cf,Es,Fm,Md,No
16.	Transition Metals 8B, 2B	Ni, Pd, Pt,Zn,Cd,Hg
17.	Soft metals II	Pb, Bi, Po, In, Tl
18.	Sulfur, Oxygen anions	S^{-2}. SO_3^{-2}, SO_4^{-2}, O^{-2}, O_2^{-}, O_2^{-2}, OH^{-}
19.	Carbon nitrogen, phosphorous based anions	NO_3^{-}, NO_2^{-}, N^{-3}, CO_3^{-2}, C^{-4}, C_2^{-2}, PO_4^{-3}
20.	Halogen based oxyanions	ClO_4^{-},ClO_3^{-},$ClO_2^{-}$$ClO^{-}$, BrO_3^{-}, IO_3^{-}
21.	Metal and metalloid based anions	$Al(OH)_4^{-}$,MnO_4^{-}, CrO_4^{-2},$Cr_2O_7^{-2}$, AsO_4^{-3}
22.	Halides (ask halide chemistry specific questions, different from element specific in group 10) and ammonium	F^{-}, Cl^{-}, Br^{-}, I^{-}, NH_4^{+}

NOTE: This exercise has been used in the first semester of general chemistry. The qualitative schemes are written in the first week of the assignment. Students are given all qualitative schemes to solve in the second week of the assignment.

REFERENCES

[1]   Bailar, Jr., J. C.; Moeller, T.; Kleiberg, J.; Guss, C. O.; Castellion, M. E.; Metz, C. *Chemistry with Inorganic Qualitative Analysis*, 3rd ed., Harcourt Brace Jovanovich: New York. NY, **1989**.

[2]   Lide, D. R. *CRC Handbook Of Chemistry And Physics*, 89th ed., CRC Press, **2008**.

Send Orders of Reprints at reprints@benthamscience.net

CHAPTER 5

Molecular Geometries, Hybridizations and Polarities

Thomas J. Manning[*] and Aurora P. Gramatges

Department of Chemistry, Valdosta State University, Valdosta, Georgia, USA, and Instituto Superior de Tecnología y Ciencias Aplicadas, La Habana, Cuba

Abstract: In this exercise, the basics of Valence Shell Electron Pair Repulsion (VSEPR) will be reviewed and the structures constructed using this approach will be visualized in three dimensions. Students will build molecules with geometries such as linear, trigonal planar, tetrahedral, trigonal bipyramidal, square planar, and octahedral. Students will identify the hybridizations of the various molecules constructed in the molecular modeling program. Students will utilize semiempirical methods to calculate the dipole moments of the molecular species constructed.

Keywords: Geometry, VSEPR theory, hybridization, molecular modeling, semiempirical methods, dipole moment.

INTRODUCTION

In this exercise the student will construct a series of molecules with sp, sp^2, sp^3, sp^3d and sp^3d^2 hybridizations. They will than perform semi empirical calculations (PM3) on each structure and use the resulting data to obtain structural data on its geometry (bond angles, bond lengths) as well as their dipole moments. As with all of these exercises, it is assumed that students have access to a general chemistry textbook and are familiar with specific topics related to molecular geometries [1,2]. This exercise assumes that students have some background on constructing a structure using Valence Shell Electron Pair Repulsion (VSEPR), and have been exposed to concepts such as hybridization and polarity.

The notation AB_xL_y represents the central atom (A) which is the atom we examine when determining hybridization, B is the number of bonds protruding from the central atom. A single, double or triple bond all count as one in this number. And

---

*Address correspondence to Thomas J. Manning: Department of Chemistry, Valdosta State University Valdosta GA 31698, USA; Tel: 229-333-7178; E-mail: tmanning@valdosta.edu

L is the number of lone pairs entered only on the central atom. For example, methane (CH_4) would be AB_4L_0 or AB_4 because the carbon only has four single bonds (C-H) and no lone pairs. We typically omit the letter if the subscript is zero. Likewise CCl_4 would be AB_4 because carbon still only has 4 single bonds and no lone pairs. The lone pairs on the chloride ions (-Cl) do not count. Water (H_2O) would be AB_2L_2 because the oxygen atom, which is the central atom, has two bonds and two lone pairs.

From the previous lab, we assume the students are familiar with some of the basics of using the Spartan software. This exercise will take students through the construction, calculations and evaluations of six structures: $BeCl_2$, $SnCl_2$, C_2H_2, XeF_4, I_3^-, SF_6 and then allow them to work with an additional twenty-two structures. The three dimensional images coupled with the computational results should provide clear images for the various geometries and their polarities routinely encountered by chemists.

Pre-Lab Exercise

In the exercise below there are twenty eight molecules. Construct the Lewis structures on scrap paper (bonds and lone pairs). A reference source (textbook, website) that assigns molecular and electronic geometries for the hybridizations listed (sp, sp^2, sp^3, dsp^3, d^2sp^3) can be used.

Table 1: The number of valence electrons for elements that may be encountered in this exercise.

Element	H	Be	B	C, Si, Sn	N,P	O,S	F,Cl,Br,I	Ne, Ar, Kr, Xe
# valence Electrons	1	2	3	4	5	6	7	8

First, set up a table with seven columns across and twenty-nine rows down and label it as shown in Table **2**. In your report it is recommended that it be in Portrait format. It is recommended to work out the Lewis Structures (bonds, lone pairs) on scrap paper and then construct them on the computer. Use a 2D drawing program to construct the flat structures for your 29 structures. Be sure to enter your name, date, lab title and instructor at the top of the first page.

1. $BeCl_2$: The first molecule to be studied in Spartan is $BeCl_2$. **Be** is the central atom and is the key atom for determining hybridizations and geometries. The chloride ions are attached to the central atom and their bonds and lone pairs will follow that outlined in Table **1** from Chapter 2. In all of the structures you construct in this exercise the atoms that are linked to the central atom will follow the rules outlined in Table **1** from Chapter 2 but the bonds and lone pairs on the central atom will be determined on a structure by structure basis.

1. Be donates two valence electrons and each chloride donates seven valence electrons for a total of sixteen (16) valence electrons. Two electrons, as either a single bond or a lone pair, are represented by a single line. Following Table **1** from Chapter 2, we place a single bond and three lone pairs around each chlorine atom and place Be in the middle. Adding up the electrons totals sixteen. Using a reference source (*i.e.* textbook), this structure qualifies as a "linear" geometry (AB_2) with sp hybridization.

In Spartan we will now build this structure. Be sure all other structures are closed and open a new page. Under the "exp" tab, select "Be", chose the linear structure "-*-", the single bond "–" and insert the Be atom in the workspace. Now select the hydrogen atom from the "Ent" (or "Exp") page. Be sure you select the "*-" if working in the "Exp". Go to "build" tab and then "minimize" your structures energy. Be sure to save it with a unique, recognizable name in two locations (including an external device). Once saved, go to "Set Up" and "Calculations" and select the following:

1. Single Point energy

2. "Semi empirical" and "PM3"

3. Start from "Initial" geometry

4. Check "Symmetry"

5. Total Charge "Neutral"

6. Compute "El. Charges"

7. Multiplicity "Singlet"

8.   Under Print click "Atomic Charges".

9.   Click in "Converge".

10. Click "Global Calculations".

11. Click "Submit" and then "Ok" when it asks if you have started the run and "Ok" when it has completed.

12. Be sure to make the background white and enlarge the structure so it occupies the whole work area.

13. Measure the bond angle with the central atom as the second element. Measure the bond angles for any bonds involving the central atom.

14. Click on "Display" and "Properties" and record your dipole moment.

2. SnCl$_2$: The second molecule to be constructed is tin (II) chloride. On a periodic table note that tin is below carbon so it has 4 valence electrons. Each chlorine atom will have seven valence electrons for a total of eighteen valence electrons between the three atoms. Using Table **1** from Chapter 2, each chlorine atom is assigned one bond and three lone pairs of electrons, which accounts for sixteen of the eighteen electrons. This leaves one pair of electrons that were not used and are assigned to the central atom (Sn) as a lone pair of electrons. This results in a central atom with a sp^2 hybridization (AB$_2$L$_1$ or AB$_2$L).

Be sure to save and close your last Spartan file and open a new one. Select "Sn" from the "Exp" table and select the bent structure. Select "Cl" and use the single bond option "$_*$-". Minimize the structures energy and save it. Follow the same computational procedure described in the BeCl$_2$ computations. Copy the structure and record the specific values outlined into your report.

3. C$_2$H$_2$: Ethyne or acetylene has a total of ten valence electrons, four from each carbon and one from each hydrogen atom. Each carbon atom functions as a central atom and hydrogen has a single bond and no lone pairs. The resulting structure involves a single triple bond between the two carbon atoms and two single C-H bonds. This is a sp hybridized structure. Comparing it to $BeCl_2$ we see that both structures are sp hybridized but the $BeCl_2$ has only single bonds while the C_2H_2 has a triple and a single bond. It is not the type of bond (single, double, triple) protruding from the central atom that dictates the hybridization but the number of bonds and lone pairs on the central atom (s).

In Spartan we will now build this structure. In Spartan ethyne requires the selection of a C atom that has a single bond and a triple bond. This bonding sequence can be found on the "Ent" page. Linking the carbon atoms together by the triple leaves two single bonds for the attachment of the hydrogen atoms. Once constructed, follow the computational approach outlined above and save, copy and paste your structure and the data into your report.

4. XeF_4: The fourth molecule to be constructed is xenon tetrafluoride. Xe, an inert gas, contributes eight valence electrons to the Lewis structure. Fluorine contributes seven for each atom or twenty-eight electrons total. In drawing the structure there are thirty-six valence electrons and each fluorine atom will have one bond and three lone pairs or account for a total of thirty-two valence electrons ($8x4 = 32$). This leaves a total of 4 electrons ($36-32 = 4$) that are assigned to the central atom as two lone pairs of electrons. This means Xe, the atom that determines the molecules geometry, will have four single bonds and two lone pairs protruding from it.

Xenon tetrafluoride has a square planar geometry, which can be found on the "Exp" page. In addition to selecting the square planar selection, select single bond. Fluorine can be selected from either page and is attached to the four single bonds protruding from Xe. Minimize the structure and set up the calculations as outlined above. Two angles can be measured with this symmetric structure, $90°$ and $180°$.

Some versions of Spartan may not be able to run this structure using Semiempirical methods. If this is the case, use "Molecular mechanics" and "MMFF". (If you encounter other structures with larger atoms that Spartan cannot handle in semi empirical, also use MM). after calculations are complete, record the parameters in report and copy the structure also.

5. I_3^-: The fifth molecule to be built and inserted in your report is I_3^-. Each iodine atom will donate seven valence electrons for a total of twenty-one electrons. In most cases you will have an even number of electrons to distribute so always check your calculations if you come up with an odd number. In this case the triatomic molecule is also an anion which means you add an additional electron for a total of twenty-two valence electrons. One of the iodine atoms is the central atom and the other two are connected to it (*i.e.* I-I-I). The two attached atoms will have one bond to the central atom and three lone pairs. Between the two iodine atoms, this accounts for 16 electrons, leaving 6 electrons or three pairs unaccounted for (22 valence – 16 on I's = 6 left over). These three lone pairs are attached to the central iodine atom making it an AB_2L_3 with a sp^3d (or dsp^3) hybridization.

In constructing I_3^-, there are two new aspects to building and running the structure. First, when you select the first iodine atom (central atom) use the "Exp" page and select Iodine with "-*-" or two bonds and select them to be single bonds. Next, go to the "ent" page and select iodine again and connect them to the two dangling bonds on the central atom and minimize the structure. In the calculations set up, change Total Charge to "anion", because your I_3^- structure has a -1 charge. In future structures that are positively or negatively charge, be sure to consider the charge. Click on "Model" and "label" and note which iodine atoms are numbered 1, 2, and 3. Now go to "Display" and "Output". Scroll down and you'll see a calculation result called "Atomic charges" and notice that each iodine atom has a partial negative charge. This is the average distribution of the -1 charge over the molecule. You'll also notice that while the molecule does not have a net dipole moment, it does have charges on each atom. Again, make the required measurements and move the data to your report.

6. SF_6: The sixth and final structure that will be outlined is sulfur hexafluoride. Sulfur will contribute six valence electrons and each fluorine atom will contribute seven for a total of forty-eight electron ($1xS(6\ e^-) + 6\ x\ F(7e^-) = 48e^-$). Following Table **1** from Chapter 2, each fluorine atom will have one bond and three lone pairs for a total of eight electrons per fluorine atom. Considering six fluorine atoms, this accounts for all forty eight valence electrons being tied up in six single bonds and eighteen (6F x 3) lone pairs resulting in an octahedral geometry.

Save and close any open structures and open a new workspace. Sulfur hexafluoride will now be constructed. Select S from the "Exp" page and then select the octahedral structure (it has 6 bonds protruding). On the "Ent" page select the fluorine atom and connect it to the six dangling bonds on the central atom and minimize the structure. Like XeF_4 above, if this structure does not work with Semiempirical methods, run it on Molecular mechanics. Also, under "Display" and "Output" you'll find that the dipole moment is listed for the molecule. If you experiment with different variations in calculations (*i.e.* change Single point Energy to Equilibrium geometry; or run in Molecular mechanics and Semiempirical) you may see a small shift in bond distances.

Table 2: Set up your table, in Portrait orientation, in the following format. Instructions and key points for the first six molecules are provided above. Your table should have seven columns across and twenty nine rows down. Use the same headers in your table that are listed below (Hybrid=hybridization; Molee=Molecular). Several of the structures are completed and inserted in the table.

Species	AB_xL_y	Hybrid., Electronic geometry	Molec. Geom.	VSEPR 2-D structure	Spartan Structure	Properties Calculated
$BeCl_2$	AB_2	sp linear	linear	$|\overline{Cl}-Be-\overline{Cl}|$		Angle=180° Be-Cl (2), 1.36 Å Debye (nonpolar)
$SnCl_2$	AB_2L	sp^2 trigonal planar	bent	$|\overline{Cl}\diagup^{\overline{Sn}}\diagdown\overline{Cl}|$		Angle = 109° Cl-Sn=2.32 Å 4.11 Debye (polar)
C_2H_2	AB_2	Sp linear	linear	H−C≡C−H		Angle =180° C-H 1.066 Å CC, 1.20 Å 0.0 Debye (nonpolar)
XeF_4	AB_4L_2	d^2sp^3 octahedral	Square planar	F\\\\ F—Xe—F \\\\ F		Angles=90°,180° Xe-F, 1.924 Å 0.0 Debye (nonpolar)

I_3^-	AB_2L_3	dsp^3 trigonal bipyramidal	Linear	$|-I-I-I|$		Angle = 180° I-I=1.926 Å 0.0 Debye (nonpolar)
SF_6	AB_6	d^2sp^3 octahedral	octahedral			Angles = 180°, 90° S-F = 1.660 Å 0.0 Debye (nonpolar)

Your report should include the six molecules completed above using data you've obtained. You will add an additional twenty-two structures, some common and some obscure, to help you better understand and visualize various geometries and hybridizations and their impacts on bond angles and distances as well as dipole moments. Next to some structures is a hint about construction. These structures are:

- SF_5I (sulfur is the central atom, how does its dipole moment compare to SF_6?)

- H_2O, (with two single bonds, oxygen is always bent, NOT -*- but bent on Exp page).

- NH_3, (with nitrogen in the middle, its geometry looks like a tripod stand).

- CH_4, (a classic tetrahedral molecule, note the changes in dipole moment as you replace H with Cl).

- CH_3Cl, (chloromethane).

- CH_2Cl_2, (dichloromethane).

- $CHCl_3$, (trichloromethane).

- CCl_4, (tetrachloromethane).

- C_2H_4, (unsaturated molecule).

- C_2H_6, (saturated molecule, compare C_2H_2, C_2H_4 and C_2H_6 geometries in 3-D).

- H_2S, (S is under O in periodic table so this molecule structurally looks like water).

- H_2SO_4, (S is in center, O's will have either 2 single bonds (S-O-H, select bent O) or a double bond (S=O).

- H_3PO_4, (P is in center, O's will have either 2 single bonds (P-O-H, select bent) or a double bond (P=O).

- SF_4.

- BrF_3.

- IF_5.

- ICl_3.

- H_3O^+ (instead of checking neutral in Spartan, check cation, O has three bonds, like a tripod).

- NH_4^+ (check cation in Spartan, this will have a tetrahedral shape).

- $TeCl_4$ (if semi empirical does not run, use molecular mechanics).

- H_2O_2 (each oxygen has two bonds, both bent).

- $XeCl_2F_2$ (construct two structures, one in which Cl's and F's are next to each other, and one where Cl and F are opposite, see Fig. **1**).

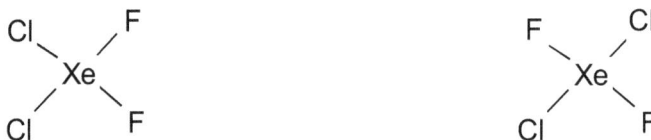

Figure 1: The right structure is referred to as a "cis" structure and the left is referred to as a "trans" structure.

Post Lab Questions

1. Why do we only consider the central atom when assigning hybridizations and geometries?

2. For the series CH_4, CH_3Cl, CH_2Cl_2, $CHCl_3$, CCl_4 discuss what impact the replacement of hydrogen by chlorine atoms had on the overall dipole moment of the molecule.

3. Molecules such as I_3^- and NH_4^+ are symmetric and have no dipole moment making them nonpolar. Why are they water soluble?

4. For acetic acid (CH_3COOH), do both carbon atoms have the same geometry? Hybridization? (see Fig. **2**)

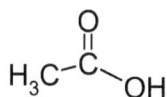

Figure 2: Acetic acid (HAc, CH_3COOH, $C_2H_3O_2$) is a common weak acid.

REFERENCES

[1]   Atkins, P. W.; De Paula, J. *Atkins' Physical Chemistry*, 9th ed., Oxford University Press, **2009**.

[2]   Atkins, P. W.; Friedman, R. S. *Molecular Quantum Mechanics*, 5th ed., Oxford University Press, **2010.**

CHAPTER 6

Twenty-Five Periodic Trends. Graphing the Periodic Table to Life

Thomas J. Manning[*] and Aurora P. Gramatges

Department of Chemistry, Valdosta State University, Valdosta, Georgia, USA, and Instituto Superior de Tecnología y Ciencias Aplicadas, La Habana, Cuba

Abstract: In this exercise students will learn chemical and physical properties associated with specific elements, and will learn how to graph a number of trends as a function of different groups of elements and determine if a periodic trend exist.

Keywords: Periodic trends, periodic table, chemical and physical properties, correlations.

INTRODUCTION

The periodic table represents many chemical and physical trends [1]. In this exercise the student will use a spreadsheet to graph a total of thirty potential trends and provide a brief explanation (2-3 sentences) for each trend. Tables **1 - 4** contain chemical and physical parameters for four different sets of elements [2]. Students will use the data in these tables to begin to explore periodic trends that may or may not exist. Table **5** represents the twenty-five correlations that students are required to graph in Excel. The data will be entered in the spreadsheet and graphed in a two dimensional plot. Both axes should be labeled and a chart title and your name will be included on each graph. Also, a figure caption should describe each graph, and a best fit line and a correlation coefficient should be visible on the graph. Fig. **1** provides an example of the graph and figure caption. Each page (8.5 x 11) should contain three graphs and captions.

In some cases there may NOT be a good correlation so note this in your explanation. Once the twenty-five graphs are complete, the student will pick five additional trends that have not yet been plotted and include them (**26-30** in

---

*Address correspondence to Thomas J. Manning:** Department of Chemistry, Valdosta State University Valdosta GA 31698, USA; Tel: 229-333-7178; E-mail: tmanning@valdosta.edu

your report). For **26-30**, there should be at least one plot from each table (*i.e.* Tables **1-4**). With three plots per page and thirty plots total, your report should be ten pages long – exactly! There are no pre or post lab questions associated with this exercise.

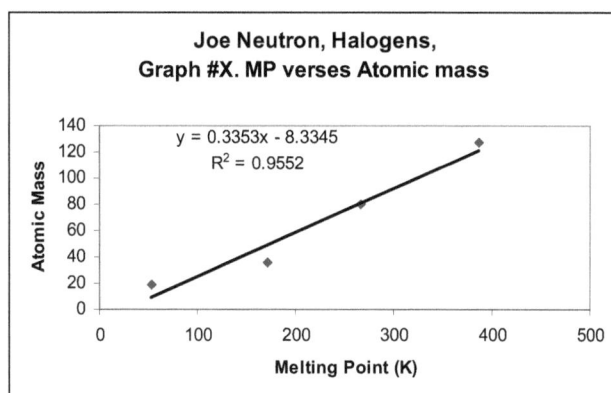

**Joe Neutron, Halogens,
Graph #X. MP verses Atomic mass**

$y = 0.3353x - 8.3345$
$R^2 = 0.9552$

Figure 1: The atomic mass of the halogen atoms (F, Cl, Br, I) is plotted against the melting point of the substances, which are diatomic molecules in the solid and liquid phase (*i.e.* I_2, Br_2, *etc.*). As the mass of the atoms increase, the melting point increases also. If unfamiliar with the Best Fit (linear or exponential) or the correlation coefficient (R^2) options in Excel, use the Help option in the upper right hand corner.

This exercise assumes that students have completed previous spreadsheet exercises and is familiar with data graphing. While there are no pre or post lab questions, students may have to familiarize themselves with the physical or chemical concepts listed in the tables (electronegativity, heat of fusion, *etc.*) [2,3].

Table 1: Some physical and chemical parameters associated with the halogens.

Element	State[1]	BP[2]	MP[3]	M[4]	AR[5]	ZN[6]	RP[7]	S[8]	TC[9]	EC[10]	D[11]	IP[12]	EN[13]	HF[14]
Fluorine (F$_2$)	G	85.1K	53.63K	18.9984	0.57Å	9	2.87	202	0.000279	_	1.696 g/L	17.422	3.98	-
Chlorine (Cl$_2$)	G	239.25K	172.31K	35.4527	0.97Å	17	1.36	223	0.000089	_	3.214 g/L	12.967	3.16	-
Bromine (Br$_2$)	L	332.4K	266.05K	79.904	1.12Å	35	1.08	152	0.0012	_	3.119 g/cc	11.814	2.96	-
Iodine (I$_2$)	S	458.55K	386.65K	126.9045	1.32Å	53	.535	116	0.00449	-	4.93 g/cc	10.451	2.66	-

[1]State, s=solid, l=liquid, g=gas.
[2]BP = Boiling point in Kelvin.
[3]MP = Melting point in Kelvin,
[4]M = Atomic mass (g/mole)
[5]AR = Atomic radius (Å)

[6]ZN = Z [#] (# of protons)
[7]RP = Standard Reduction Potential (Volts)
[8]S = Entropy in Joules/Kelvin.mol
[9]TC = Thermal conductivity, W/cmK
[10]EC = Electrical conductivity 10^6/cmΩ
[11]D = Density (note units and phase! Convert cc or cm^3 to liters)
[12]IP = Ionization Potential (eV, first)
[13] EN = Electronegativity
[14]HF = Heat of Fusion (kJ/mol)

Table 2: Some physical and chemical parameters associated with the alkali metals.

Element	State[1]	BP[2]	MP[3]	M[4]	AR[5]	ZN[6]	RP[7]	S[8]	TC[9]	EC[10]	D[11]	IP[12]	EN[13]	HF[14]
Lithium	Solid	1615	453.85	6.941	2.05	3	-3.05	29.1	.847	.108	.534	5.392	.98	2.09
Sodium	Solid	1156	96.96	22.98	2.23	11	-2.71	51.2	1.41	.21	.971	5.139	.93	2.59
Potassium	Solid	1032	336.5	39.09	2.77	19	-2.93	64.6	1.024	.139	.862	4.341	.82	2.33
Rubidium	Solid	961	312.7	85.46	2.98	37	-2.98		.582	.0779	1.63	4.177	.82	2.19
Cesium	Solid	944	301.7	132.9	3.34	55	-2.92	85.2	.359	.0489	1.87	3.894	.79	2.092

[1]State, s=solid, l=liquid, g=gas.
[2]BP = Boiling point in Kelvin.
[3]MP = Melting point in Kelvin,
[4]M = Atomic mass (g/mole)
[5]AR = Atomic radius (Å)
[6]ZN = Z # (# of protons)
[7]RP = Reduction Potential (Volts)
[8]S = Entropy in Joules/Kelvin.mol
[9]TC = Thermal conductivity in W/cmK
[10]EC = Electrical conductivity 10^6/cmΩ
[11]D = Density (g/cm^3)
[12]IP = Ionization Potential (eV, first)
[13] EN = Electronegativity
[14]HF = Heat of Fusion, kJ/mole

Table 3: Some physical and chemical parameters associated with the s,d,p block elements.

Element	State[1]	BP[2]	MP[3]	M[4]	AR[5]	ZN[6]	RP[7]	S[8]	TC[9]	EC[10]	D[11]	IP[12]	EN[13]	HF[14]
Potassium	S	1032	3365	39.0983	2.77	19	-		1.024	.139	.862	4.341	.82	2.334
Calcium	S	1757	1112	40.078	2.23	20	-		2.01	.298	1.55	6.113	1	8.54
Scandium	S	3104	1812	44.95591	2.09	21	-		.158	.0177	2.99	6.54	1.36	14.1
Titanium	S	3560	1933	47.88	2	22	-		.219	.0234	4.54	6.82	1.54	15.45
Vanadium	S	3682	2175	80.9415	1.92	23	-		.307	.0489	6.11	6.74	1.63	20.9
Chromium	S	2945	2130	51.9961	1.85	24	-		.937	.0774	7.19	6.766	1.66	16.9
Manganese	S	2235	1517	54.93805	1.79	25	-		.0782	.00695	7.43	7.435	1.55	12.05
Iron	S	3023	1808	55.847	1.72	26	-		.802	.0993	7.874	7.87	1.83	13.8
Cobalt	S	3143	1768	58.9332	1.67	27	-		1	.0172	8.9	7.86	1.88	16.19
Nickel	S	3005	1726	58.6934	1.62	28	-	29.87	.907	.143	8.9	7.635	1.91	17.47
Copper	S	2840	1357.75	63.546	1.57	29	-		4.01	0.596	8.96	7.726	1.9	13.05
Zinc	S	1180	692.88	65.39	1.53	30	-		1.16	0.166	7.13	9.394	1.65	7.322

Table 3: contd...

Gallium	S	2676	303.05	69.723	1.81	31	-		0.406	0.0678	5.907	5.99	1.81	5.59
Germanium	S	3103	1210.55	72.61	1.52	32	-		0.599	1.45E-8	5.323	7.899	2.01	36.94
Arsenic	S	876	1081	74.922	1.33	33	-		0.502	0.0345	5.72	9.81	2.18	369.9
Selenium	S	958	494	78.96	1.22	34	-		.0204	1.0E-12	4.79	9.752	2.55	6.694
Bromine	L	332.4	266.05	79.904	1.12	35	-		.00122	0	3.119	11.814	2.96	5.286
Krypton	G	119.95	115.93	83.8	1.03	36	-		.0000949	0	3.75	13.99	2.94	1.638

[1]State, s=solid, l=liquid, g=gas at 25 °C and 1 atm.
[2]BP = Boiling point in Kelvin.
[3]MP = Melting point in Kelvin,
[4]M = Atomic mass (g/mole)
[5]AR = Atomic radius (Å)
[6]ZN = Z # (# of protons)
[7]RP = Reduction Potential (Volts). Because elements have different oxidation states (K^+, Ca^{+2}, *etc.*), not considered here.
[8]S = Entropy in Joules/Kelvin.mol
[9]TC = Thermal conductivity in W/cmK
[10]EC = Electrical conductivity 10^6/cmΩ
[11]D = Density (g/cm^3)
[12]IP = Ionization Potential (eV, first)
[13]EN = Electronegativity
[14]HF = Heat of Fusion, kJ/mole

Table 4: Some physical and chemical parameters associated with some p-block elements.

Element	State[1]	BP[2]	MP[3]	M[4]	AR[5]	ZN[6]	HV[7]	S[8]	TC[9]	EC[10]	D[11]	IP[12]	EN[13]	HF[14]
Carbon (graphite)	S	5100	3773	12.011	.91	6	355.8	5.6	1.29	.00061	2.26	11.26	2.55	104.6
Nitrogen	G	77.5	63.29	14.00674	.75	7	2.79	191	.0002598	0	1.2506	14.534	3.04	.3604
Silicon	S	2628	1683	28.0855	1.46	14	439	18.8	1.48	2.52 x 10^{-12}	2.33	8.151	1.9	50.55
Phosphorous (white)	S	553	317.45	30.97376	1.23	15	12.43	41.1	.000235	10^{-17}	1.82	10.486	2.19	.657
Germanium	S	3103	1210.55	72.61	1.52	32	330.9		.599	1.45 x 10^{-8}	5.323	7.899	2.01	36.94
Arsenic	S	876	1081	74.92159	1.33	33	34.76		.502	.0345	5.72	9.81	2.18	369.9
Tin	S	2543	505.21	118.71	1.72	50	290.4	51.1	.666	.0917	7.31	7.344	1.96	7.029
Antimony	S	1860	904.05	121.757	1.53	51	77.14		.243	.0288	6.684	8.641	2.05	19.87
Lead	S	2013	600.75	207.2	1.81	82	179.4	64.8	.353	.0481	11.35	7.416	2.33	4.799
Bismuth	S	1837	544.67	208.9804	1.03	83	104.8		.0787	.00867	9.75	7.289	2.02	11.3

[1]State, s=solid, l=liquid, g=gas at 25 °C and 1 atm.
[2]BP = Boiling point in Kelvin.
[3]MP = Melting point in Kelvin,
[4]M = Atomic mass (g/mole)
[5]AR = Atomic radius (Å)
[6]ZN = Z # (# of protons)
[7]HV = Heat of vaporization (kJ/mol)
[8]S = Entropy in Joules/Kelvin.mol

[9]TC = Thermal conductivity in W/cmK
[10]EC = Electrical conductivity 10^6/cmΩ
[11]D = Density (g/cm^3)
[12]IP = Ionization Potential (eV, first)
[13] EN = Electronegativity
[14]HF = Heat of Fusion, kJ/mole

Table 5: Students are required to plot the following graphs and provide a brief explanation of each observation.

Table # (above) and Periodic group	Trend to be graphed. X-axis (first) and Y-axis (second)	Numbers
1 (Halogens)	Atomic radius verses Melting Point	5 *vs.* 13
1 (Halogens)	Reduction Potential verses Electronegativity	7 *vs.* 13
1 (Halogens)	Atomic radius verses Ionization Potential	5 *vs.* 12
1 (Halogens)	Boiling Point verses Melting Point	2 *vs.* 3
1 (Halogens)	Molar mass verses Boiling Point	2 *vs.* 4
1 (Halogens)	Atomic radius verses Reduction Potential	5 *vs.* 7
1 (Halogens)	Atomic radius verses Electronegativity	5 *vs.* 13
2 (Alkali metals)	Electronegativity *vs.* Ionization Potential	12 *vs.* 13
2 (Alkali metals)	Electrical conductivity verses thermal Conductivity	9 *vs.* 10
2 (Alkali metals)	Electrical Conductivity *vs.* melting Point	3 *vs.* 10
2 (Alkali metals)	# Protons verses Atomic Radius	5 *vs.* 6
2 (Alkali metals)	Melting Point verses Heat of Fusion	3 *vs.* 14
2 (Alkali metals)	Ionization Potential verses Atomic Radius	5 *vs.* 12
3 (K-Kr)	Thermal Conductivity verses Electrical Conductivity	9 *vs.* 10
3 (K-Kr)	Atomic Radius verses Ionization Potential	5 *vs.* 12
3 (K-Kr)	# Protons verses Atomic Radius	5 *vs.* 6
3 (K-Kr)	Heat of Fusion verses Melting Point	3 *vs.* 14
3 (K-Kr)	Boiling Point verses Electrical Conductivity	2 *vs.* 10
3 (K-Kr)	# of Protons verses Density	6 *vs.* 11
4 (P-Block)	Electrical Conductivity verses Thermal Conductivity	9 *vs.* 10
4 (P-Block)	Ionization Potential verses Electronegativity	12 *vs.* 13
4 (P-Block)	Boiling Point verses # of Protons	2 *vs.* 6
4 (P-Block)	Melting Point verses Atomic Radius	3 *vs.* 5
4 (P-Block)	Density verses Boiling Point	2 *vs.* 11
4 (P-Block)	Density verses Entropy	8 *vs.* 11

REFERENCES

[1]   Dynamic Periodic Table http://www.ptable.com/?lang=en (accessed June 16, **2012**).

[2]   Lide, D. R. *CRC Handbook Of Chemistry And Physics*, 89th ed., CRC Press, **2008**.

[3]   Petrucci, R. H.; Harwood, W. S.; Madura, J. D. *General chemistry: principles and modern applications*, Pearson/Prentice Hall, **2007**.

CHAPTER 7

Titration Involving a Strong Base and Weak Acid

Thomas J. Manning* and Aurora P. Gramatges

Department of Chemistry, Valdosta State University, Valdosta, Georgia, USA, and Instituto Superior de Tecnología y Ciencias Aplicadas, La Habana, Cuba

Abstract: In this exercise students will learn about the reactions between a strong base and weak, as well as a weak base and a strong acid. Students will simulate a titration curve that would be obtained if a strong acid and a weak base were reacted, and they will advance their knowledge of using a spreadsheet in chemistry related exercises.

Keywords: Strong base, weak acid, titrations, titration curves, equilibrium constant.

INTRODUCTION

In a previous lab (see Chapter #3) the titration of a strong acid (SA) and a strong base (SB) was simulated in a spreadsheet exercise. The SA/SB titrations involve the formation of water and spectator ions (*i.e.* Na^+, Cl^-) [1]. Acid/base spectator ions such as Na^+, K^+, Cl^- and NO_3^- have a negligible impact on the pH. When we replace the strong acid with a weak acid or we replace the strong base with a weak base the simulation becomes more complicated because of the equilibrium constants involved (K_a, K_b) [2,3].

In this titration simulation, sodium hydroxide is in the beaker (0.05 M, 50 mLs) and the acetic acid (HAc) is in the buret (0.075 M) and will be titrated in 0.5 mL increments until 100 mLs of the weak acid have been delivered. Acetic Acid ($HC_2H_3O_2$, CH_3COOH) and sodium hydroxide reaction follows the equation,

$$HAc(aq) + NaOH(aq) ==> H_2O(l) + Ac^-(aq) + Na^+(aq)$$

While water is neutral and $Na^+(aq)$ is a spectator ion in acid/base reactions, Ac^- (aq) or the acetate ion is a weak base.

---

*Address correspondence to Thomas J. Manning: Department of Chemistry, Valdosta State University Valdosta GA 31698, USA; Tel: 229-333-7178; E-mail: tmanning@valdosta.edu

$$Ac^-(aq) + H_2O(l) ==> HAc(aq) + OH^-(aq) \ K_b = 5.5 \times 10^{-10}$$

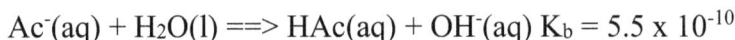

And the hydroxide concentration can be estimated from:

$$[OH^-] = (K_b*[Ac\text{-}])^{1/2} ==> (5.5 \times 10^{-10} [Ac^-])^{1/2}$$

This expression can be used when the concentration of the acetate is much greater (1000 X) than the K_b. When you calculate the [Ac⁻] after each addition you can also estimate the [OH⁻] from the acetate and confirm that it is much less than the [OH⁻] from the NaOH. Acetic acid in water is a weak acid, producing hydronium and acetate,

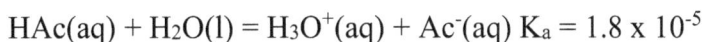

$$HAc(aq) + H_2O(l) = H_3O^+(aq) + Ac^-(aq) \ K_a = 1.8 \times 10^{-5}$$

And the hydronium concentration ($[H_3O^+]$) can be estimated/approximated from

$$[H_3O^+] = (K_a*[HAc])^{1/2} = (1.8 \times 10^{-5}*[HAc])^{1/2}$$

Like the [OH⁻] calculation above, this equation can be used to estimate the hydronium ion concentration when the weak acid concentration (HAc) is much greater (*i.e.* 1000 X) than the K_a. In this simulation you will encounter two regions. The first region is when there is excess hydroxide, or

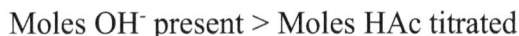

Moles OH⁻ present > Moles HAc titrated

While the reaction does produce Ac⁻, a weak base, the hydroxide from the strong electrolyte NaOH will control the pH. The amount of OH⁻ produced by the Ac⁻ anion is considered negligible compared to the NaOH present in these calculations and can be confirmed using equation.

The second region on the titration curve will take place after the equivalence point or when more moles of acetic acid have been titrated than moles of sodium hydroxide were originally in the beaker. In this region the excess acetic acid will control the pH.

The equivalence point for the strong acid and strong base titration simulated will in exercise 3 take place at a pH of seven (7). When the titration involves a weak acid or a weak base it will shift the equivalence point away from the neutral point.

Pre-Lab Questions

Be sure to review the lab introduction outlined above. It will explain where the equations you use in the spreadsheet come from. Write out the balanced reaction and sketch the shape of the four titration curves outlined below (2 curves per page). These curves should be completed in a 2D drawing program with the axis labeled, your name and a title on top and a figure caption. It is the shape of each curve and the location of the equivalence point (pH is <, > or = to 7.0) that is important, not the exact position of the starting and ending points as these will vary with conditions and quantities of the acids and bases. Also list the reaction with the curve. Be sure to label the axis and identify where the equivalence point is in terms of acidic region or basic region. Because you don't know the concentration (assume they are in the 0.01 to 0.1 M range) or the exact volumes you can't identify the exact equivalence point, starting pH, *etc.* but your graph should represent the approximate shape observed. In each case the first species listed is in the burette and the second species listed is in the beaker.

    a.   titration of sodium hydroxide (NaOH) into acetic acid (HAc)

    b.   titration of acetic acid (HAc) into sodium hydroxide (NaOH)

    c.   titration of sodium acetate (NaAc) into hydrochloric acid (HCl)

    d.   titration of hydrochloric acid (HCl) into sodium acetate (NaAc).

Once these are complete you will begin the simulation of acetic acid (burette) being titrated into a solution of sodium hydroxide (beaker) in your spreadsheet. Your graphs, properly labeled and with a figure caption, will be transferred to your report.

Computer Exercise

    A.   Open a new spreadsheet. In box A1 place the header "Conc. of Acetic Acid". In box A2 place the number "0.075" and copy/paste it down to A202. This (0.075 M) is the concentration of acetic acid in the burette.

    B.   In box B1 type the header "Conc. of NaOH" and in box B2 enter the number ".05" and copy it down to B202. 0.05M is the starting concentration of NaOH which is in the beaker or flask.

C.   In box C1 type the header "Equilibrium Constant of HAc" and in box C2 enter the value "0.000018" (or 1.8×10^{-5}) and copy/paste it down to C202. This is the K_a for acetic acid.

D.   In box D1 enter the header "mLs of NaOH" and in D2 enter the value "50" and copy it down to D202.

E.   In box E1 enter the header "mLs of HA" and in box E2 enter the value "0". In box E3 enter the command "=SUM(E3+0.5)" and copy/paste it down to E202.

F.   In box F1 enter the header "Initial moles of HAc titrated" and in box F2 enter the formula "=SUM(A2*E2/1000)" and copy/paste it down to F202. This calculates the moles of acetic acid using the formula moles = MV. It is divided by 1000 to convert milliliters to liters.

G.   In box G1 place the header "Initial moles NaOH in beaker" and in G2 enter the equation "=SUM(D2*B2/1000" and copy/paste it down to G202. This equation is also based on moles=MV and the number (.0025) should be constant all the way down to the end of the titration.

H.   In box H1 type the header "moles of OH- after rxn" and in box H2 type the command " =IF(G2>F2,SUM(G2-F2),"" " and copy/paste it down to H202. This equation asks if there are more moles of NaOH than HAC, and if the answer is yes, it calculates the moles of OH⁻ considering the neutralization reaction in Equation 1.

I.   Type "moles HAc after neutral" in box I1 and in box I2 type the logic statement " =IF(F2>G2,SUM(F2-G2),"" ". Copy and paste this equation down to I202. This statement will determine if the moles of HAc are greater than the moles of NaOH. If they are, use the neutralization reaction in equation 1 to determine how much acetic acid is left after the reaction with hydroxide.

J.   Type the header "moles of Ac- produced" in box J1 and enter the equation " =IF(G2>H2,SUM(G2-H2),"" " in location J2. Copy and paste this equation down to J202. This statement determines if any moles of OH⁻ from NaOH are left after the neutralization reaction with acetic acid.

K. In box K1 type the label "Total Vol (Liters)" and in K2 enter the equation " =SUM((D2+E2)/1000) " and copy/paste it down to K202. This combines the 50 mls of NaOH with the HAc which has been titrated into the beaker. It is divided by 1000 to convert mLs to liters. Recall that pH and pOH must use molar concentrations of $[H_3O^+]$ and $[OH^-]$, respectively.

L. In box L1 type the header "pH when OH- is in x-s" and in K2 enter the equation " =IF(H2>0,SUM(14-LOG(H2/K2)*-1),"") " and copy/paste it down to L202. This equation determines if there is excess OH- from NaOH after the neutralization of HAc. If there is, it calculates the pOH from the equation pOH = -log[OH⁻] (LOG(H2/K2)*-1) and then uses the equality pH = 14 - pOH to determine the pH. H2 is in moles and K2 is liters resulting in a molar (moles/liter) solution of OH⁻.

M. In box M1 type the header "[H+] if HAc in x-s " and in location M2 enter the equation " =SUM((I2/K2)*C2)^(0.5) " and copy/paste it down to M202. This calculates the amount of hydronium ion $[H_3O^+]$ resulting from the dissociation of the excess acetic acid (eq. 7.5).

N. In location N1 enter the title "pH when x-s HAc" and in location N2 calculate the pH using the equation " =LOG(M2)*-1 ". Copy and paste this equation down to location N202. At this point you've completed the pH calculations for the regions where NaOH is in excess and where HAc is in excess. We have not calculated the equivalence point.

O. The equivalence point is the point in the titration where the moles of acid (HAc) equal the moles of base (NaOH). First we will use the spreadsheet and attempt to identify if we have calculated the equivalence point. In location O1 type the header "Eq. Point" and in location O2 type the equation " =IF(F2=G2," Eq. Point") " and copy/paste it down to O202. This logic statement compares the moles of OH- and HAc and prints "Eq. Point" if it is located. After you copy/paste this equation down, try to locate this statement. If it doesn't exist, visually compare columns F2 and G2 and see if there is a point where the moles of acid and base are the same. You should not an equal number of the two species. The point here is that you can conduct either a simulated or a real titration curve and NOT hit the exact equivalence point. It can be estimated or calculated from the shape of the titration curve.

P. Now we will plot the data to form a titration curve. It is assumed you've completed the SA/SB titration curve (see exercise 3), and have some familiarity with plotting data in Excel.

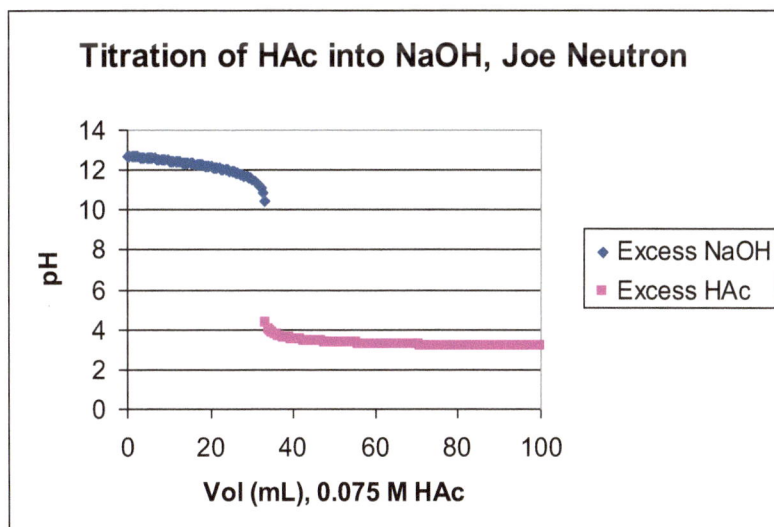

Figure 1: The simulation of the titration of a weak acid (acetic acid) into a strong base (NaOH).

Additional Exercise

Set up the titration of hydrofluoric acid (HF) into potassium hydroxide using your spreadsheet as a template. Copy and paste it into your report and list the neutralization reaction that is taking place.

REFERENCES

[1]   de Levie, R. *Aqueous Acid-Base Equilibria and Titrations*, Oxford Chemistry Primers, Oxford University Press, **1999**.

[2]   Lewis, G. N. *Valence and the Structure of Atoms and Molecules*, **1923**.

[3]   Laurence, C.; Gal, J. F. *Lewis Basicity and Affinity Scales: Data and Measurement* John Wiley & Sons., **2009**.

Send Orders of Reprints at reprints@benthamscience.net

CHAPTER 8

Modeling Weak Acids and Bases

Thomas J. Manning[*] and Aurora P. Gramatges

Department of Chemistry, Valdosta State University, Valdosta, Georgia, USA, and Instituto Superior de Tecnología y Ciencias Aplicadas, La Habana, Cuba

Abstract: In this exercise students will model the structures of monoprotic, diprotic, triprotic acids, and their conjugate bases. Students will calculate the dipole moment, molecular volume and surface area of each molecule and calculate the atomic charges on each atom in the molecules. Students will look for correlations between the acids pK_a, its dipole moment and the atomic charges on the atoms closest to the protonation/deprotonation site.

Keywords: Types of acids, polyprotic acids, equilibrium constant, molecular modeling, dipole moment.

Pre-Lab Exercise

The following questions will be answered in your report.

1. Identify the strong acids (list names and empirical formula).

2. List name/formula for a minimum of fifteen weak acids.

3. For each weak acid, provide the name and empirical formula for its conjugate base.

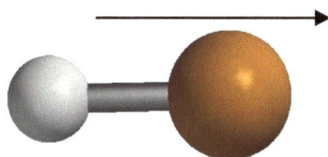

Figure 1: Hydrochloric acid has a dipole moment (1.38 D) and the hydrogen atom (+0.238) and chlorine atom (-0.238) both have partial charge or atomic charges due to the shifting of electrons over the polar molecule. The calculations are for the molecule in the gas phase.

*****Address correspondence to Thomas J. Manning:** Department of Chemistry, Valdosta State University Valdosta GA 31698, USA; Tel: 229-333-7178; E-mail: tmanning@valdosta.edu

Table **1** and Fig. **1** provide details of the type of calculations you will perform and the correlations you will search for, assuming they exists. In the data, for each molecule provided in Table **1**, notice that the positive and the negative partial charges are equal and opposite. Because each species (*i.e.* HF, HCl, *etc.*) are neutral, the partial charges on the individual atoms must add to zero. If the molecule had a negative charge (*i.e.* ClO_4^-) than the atomic charges would add to - 1. While HF is a weak acid and has a K_a, HCl, HBr and HI are strong acids and have no equilibrium for their dissociation. In this case no correlations can be drawn between K_a's (or pK_a's) and dipole moment or atomic charges. In some of the exercises below, plots such as dipole moment verses atomic charge, dipole moment verses pK_a and dipole moment verses atomic charges will be used to identify any correlations between the parameters.

Table 1: An example of correlations between dipole moment and partial charges on individual atoms.

Species	Dipole Moment	Partial Positive (H)	Partial Negative (X)
HF	1.41 D	0.327	-0.327
HCl	1.38 D	0.238	-0.238
HBr	1.26 D	0.203	-0.203
HI	0.92 D	0.149	-0.149

The computational procedures used to obtain the data in the table above were:

1. Single Point Energy, Semiempirical (PM3), Initial.

2. Check: Symmetry, Compute (Elect. Charges), Print (Atomic Charges), Converge.

3. Select: Charge (Neutral), Multiplicity (Singlet).

Students will build a series of acids and their conjugate bases or anions (typically, the anions for strong acids are not basic but spectator ions in acid/base reactions) [1-3]. Table **2** provides the format for the table that will go in your report.

Table 2: The results of calculations in the correct format for acetic acid and its conjugate base, acetate. (D = dipole moment, V = molecular volume, A = surface area). Students may find small variations in the calculated values (atomic charges, D,V, A) depending on the software version and/or computational parameters utilized. Have one table for monoprotic species, a second for diprotic species and a 3rd for triprotic species in your report.

Identifier	Name, Formula, pK	Atomic charge on four atoms	D, V, A	Structure
Monoprotic, Acid # 1	Acetic acid $HC_2H_3O_2$ $pK_a = 4.74$	-O: -0.291 =O: -0.333 -C=: 0.385	D = 4.05 D V = 61.7 Å^3 A = 83.5 Å^2	
Conjugate base # 1	Acetate $C_2H_3O_2^-$ $pK_b = 9.26$	-O: -0.648 =O: -0.643 -C=: 0.434	D = 7.02 D V = 59.12 Å^3 A = 81.06 Å^2	

In the molecular modeling software (Spartan), students should be familiar with calculating dipoles, volumes, and surface areas (see chapter 5). After building each molecule:

1. Click on Setup at the top of the page, then Calculations. There you will see a dialog box pop up. Be sure to select the following.

   Single Point Energy

   Semi-Empirical, PM3

   Initial, Symmetry (check)

   Charges (Acetic acid, neutral; Acetate, anion)

   Compute: Elect. Charges

   Multiplicity, Singlet

Print, Atomic Charges

2.   Once complete and saved, click on Display, and then Output. At the bottom of the page, you should find the title: Natural Atomic Populations and Charges. Here you will find the Atomic charges for each atom.

3.   Also, Click on Display and then Properties. Here you will find the Dipole Moment, Molecular Volume, and Surface Area.

Using Spartan, build the following structures, perform the computational procedure outlined and record the needed values into your table. Once your tables are complete, you will be instructed to perform certain plots in your spreadsheet program to search for correlations among parameters. Please note, K_a's and K_b's (pK_a's, pK_b's) are not calculated but were taken from other sources.

Table 3: Below is a list of monoprotic acids to build and evaluate. Structures are drawn without lone pairs of electrons. Lone pairs are critical in the VSEPR model when determining the geometry.

1. Hydrofluoric Acid (HF)	1. Fluoride (F⁻)
	(hint: select F with a single bond, and remove protruding bond).
2. Hydronium (H₃O⁺)	1. Water (H₂O)
  • sp³ hybridized species • tetrahedral (3 bonds, 1 lone pair) • hint, cation	 • sp³ hybridized species • 2 bonds, 2 lone pairs • Neutral

Table 3: contd....

3. Ammonium (NH_4^+) • sp^3 hybridized molecule (tetrahedral, 4 single bonds) • cation	**3. Ammonia (NH_3)** • sp^3 hybridized molecule (tetrahedral, 3 single bonds, one lone pair) • neutral molecule
4. Chloroacetic acid ($CClH_2COOH$) • sp^3 and sp^2 hybridized carbon's • Neutral molecule	**4. Chloroacetate ($C_2H_2ClO_2^-$)** • Anion (-1)
5. Nitric Acid (HNO_3, strong acid, no K_a) • Nitric acid and nitrate are delocalized structures.	**5. Nitrate (NO_3^-)** • Anion (-1) for a strong acid has minimum base properties.
6. Nitrous Acid (HNO_2, $pK_a = 3.25$) 	**6. Nitrite (NO_2^-, $pK_b = 10.75$)**
7. Hydrocyanic acid (HCN) 	**7. Cyanide (CN^-)**
8. Formic acid (CHO_2H, $pK_a = 3.75$) Formic acid is the simplest carboxylic acid. The C-O-H bond is bent (O does not have any linear structures with 2 single bonds)	**8. Formate (CHO_2^-, $pK_b = 10.25$)** Formate is the simplest carboxylate.

Table 4: The following table contains diprotic acids and the associated anions. Students should build and calculate each structure and place the results in your report.

9. Oxalic acid ($H_2C_2O_4$, pK_{a1} = 1.23) In Spartan, check "neutral" for charge.	9. Monohydrogen oxalate ($HC_2O_4^-$, pK_{a2} = 4.19) In Spartan, check "anion" for charge	9. Oxalate ($C_2O_4^{-2}$) In Spartan, check "dianion" for charge
10. Carbonic acid (H_2CO_3, pK_{a1} = 6.35) Carbonic acid is unstable and decomposes to form CO_2 and H_2O.	10. Bicarbonate (HCO_3^-, pK_{a2} = 10.33)	10. Carbonate (CO_3^{-2})
11. Sulfuric acid (H_2SO_4, strong acid) First proton dissociates 100%. Select a tetrahedral structure for sulfur from exp page.	11. Monohydrogen sulfate (HSO_4^-, pK_{a2} = 1.98) The second proton to dissociate has equilibrium constant. Select all single bonds for sulphurs tetrahedral geometry and select all single bonds for oxygen's. (not completely accurate but electrons are delocalized).	11. Sulphate (SO_4^{-2}) Select a tetrahedral geometry with all single bonds (perform cal's), than select a tetrahedral geometry with all double bonds and perform cal's. Compare your results.
12. Sulfurous acid (H_2SO_3, pK_{a1} = 1.85) (con't) Use VSEPR to confirm that the sulfur atom has three bonding areas and one lone pair for a tetrahedral geometry (select all single bonds)	12. Hydrogen sulfite (HSO_3^-, pK_{a2} = 7.20) (con't) Use the same geometry as H_2SO_3 but use a lone single bond on the oxygen atom with a -1 charge. While we draw single and double bonds with Lewis structures, this has resonance structures	12. Sulfite (SO_3^{-2}) (con't) Use same geometry as H_2SO_3 but use a lone single bond on the two oxygen atoms with a -1 charge. This is a delocalized structure and has resonance effects.

Table 5: Below is a list of triprotic acids to build and evaluate. Structures are drawn without lone pairs of electrons. Lone pairs are critical in the VSEPR model when determining a molecular geometry.

13. Phosphoric acid (H_3PO_4, pK_{a1}=2.16)	13. Dihydrogen phosphate ($H_2PO_4^-$, pK_{a2} = 7.21)	13. Monohydrogen phosphate (HPO_4^{-2}, pK_{a3} = 12.32)	13. Phosphate (PO_4^{-3})
Select a tetrahedral geometry for phosphorous with all single bonds.	Chose "anion" for charge on this species and "dianion" for the next species (HPO_4^{-2})	For the 3 structures with negative charges, each has delocalized electrons and resonance structures.	Spartan does not have an option for a tri-anion species.
14. Boric acid (H_3BO_3, pK_{a1} = 9.27)	14. Dihydrogen borate ($H_2BO_3^-$, pK_{a2} = 12.7)	14. Monohydrogen borate (HBO_3^{-2}, pK_{a3} = 13.28)	14. Borate (BO_3^{-3})
Boron, the central atom has no lone pairs resulting in a trigonal planar geometry.	Chose "anion" for this species and dianion for the next.		Spartan does not have an option for a -3 charge.

1. For your monoprotic weak acids, generate the three plots outlined below in your spreadsheet program. Be sure to label axis, place your name on the top of the graph and ONLY use a best fit line (no connect the dots). Once these graphs are complete, discuss your results (are there any correlations between the different parameters, why or why not?).

    a. Plot the atomic charge on the negative atom in the neutral species verses the dipole moment.

    b. Plot the atomic charge on the negative atom in the neutral species verses the pK_a.

    c. Plot the dipole moment verse the pK_a.

2. For your diprotic weak acids, generate the three plots outlined below in your spreadsheet program. Be sure to label axis, place your name on the top of the graph and ONLY use a best fit line (no connect the dots). Once these graphs are complete, discuss your results (are there any correlations between the different parameters, why or why not?).

   d. Plot the atomic charge on the negative atom bonded to the first proton to leave verses the dipole moment of the neutral species. (for example, in H_2CO_3, the O atom bonded to the first H to deprotonate and leave the HCO_3^-).

   e. Plot the atomic charge on the negative atom bonded to the first proton to leave verses the first pK_a (pK_{a1}).

   f. Plot the dipole moment verse the first pK_a (pK_{a1}).

Table 6: Below is a 2D Image of Glycine, An Amino Acid. Build and measure glycine, include its conjugate base, and its Zwitterion. You should have three structures in your report (species above, the zwitterions, the base or deprotonated species).

Glycine ($C_2H_5NO_2$)
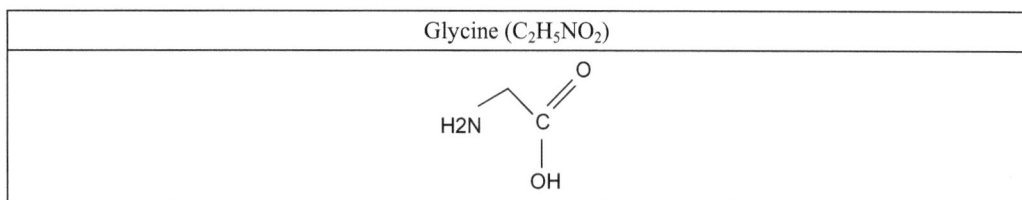

REFERENCES

[1]   de Levie, R. *Aqueous Acid-Base Equilibria and Titrations*, Oxford Chemistry Primers, Oxford University Press, **1999.**

[2]   Lewis, G. N. *Valence and the Structure of Atoms and Molecules,* The Chemical Catalog Company, Inc., **1923.**

[3]   Laurence, C.; Gal, J. F. *Lewis Basicity and Affinity Scales: Data and Measurement* John Wiley & Sons., **2009.**

CHAPTER 9

Demonstrating Bonds and Forces: From Nitrogen to Nanotubes

Thomas J. Manning* and Aurora P. Gramatges

Department of Chemistry, Valdosta State University, Valdosta, Georgia, USA, and Instituto Superior de Tecnología y Ciencias Aplicadas, La Habana, Cuba

> **Abstract:** In this exercise students will study chemical forces such as covalent bonds, ionic bonds, ion-dipole interactions and hydrogen bonds.

Keywords: Chemical bonds, interactions, nanotubes, molecular modeling, semiempirical calculations.

INTRODUCTION

Students will build a (10, 0) nanotube [1]. This relatively inert structure will be used as a template for building a sheeted peptide structure. The peptide will have the residues aspartic acid and glutamine, which are connected by a peptide bond [2]. This construction will be used to demonstrate different types of bond [3]:

a. **Covalent bonds**- chemical bonds in which atoms share electrons. For example, all the bonds in methanol (Fig. **1**), are covalent by nature.

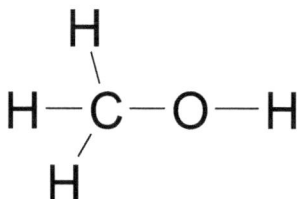

Figure 1: Methanol (CH_3OH) is a small molecule that has C-H, C-O and O-H covalent bonds.

b. **Ionic bonds** -the result of bonds between opposite oppositely charged ions. Typically the cations (+1, +2, *etc.*) and anions (-1, -2, *etc.*) that have charges greater than or equal to 1 (see Fig. **2**).

---

*Address correspondence to Thomas J. Manning: Department of Chemistry, Valdosta State University Valdosta GA 31698, USA; Tel: 229-333-7178; E-mail: tmanning@valdosta.edu

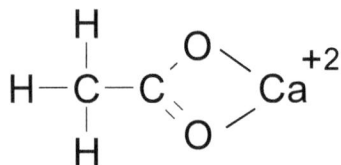

Figure 2: Acetate, with its negativity charged carboxylate, has an ionic bond with a calcium dication. The ionic bond is represented with dashed lines.

    c.   **Ion-dipole force-** is an attractive force that results from the electrostatic attraction between an ion and a neutral molecule that has a dipole moment (Fig. **3**).

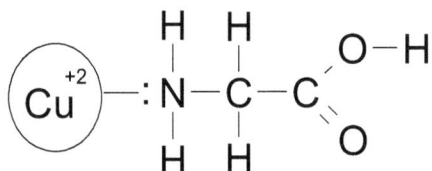

Figure 3: The attraction between copper (II) and the lone pair of electrons on the nitrogen or amine is referred to as an ion-dipole interaction.

    d.   **Dipole-dipole attraction-** exists between the dipole moments on two or more molecules. A partial positive charge or a charge less than +1 (δ^+) is attracted to a partial negative charge (δ^-) or a charge less than -1 give rise to a dipole-dipole interaction (Fig. **4**).

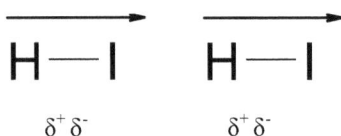

Figure 4: The partial negative charge on the iodide atom in hydroiodic acid is attracted to the partial positive charge on a hydrogen atom in another HI molecule forming a dipole-dipole interaction.

    e.   **Hydrogen bonding-** a dipole-dipole attraction that includes a hydrogen atom attached to oxygen, nitrogen, fluorine, chlorine or sulfur and attracted to a dipole on oxygen, nitrogen, fluorine, chlorine or sulfur. Because of its strength relative to other dipole-dipole interactions and its importance in biochemical and environmental systems, it is given its own classification (see Fig. **5**).

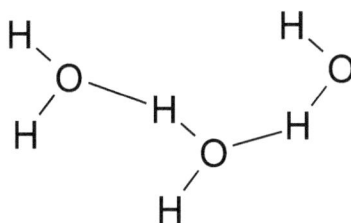

Figure 5: The partial negative charges on oxygen (δ^-) and the partial positive charges on hydrogen (δ^-) on different water molecules are hydrogen bonds.

f. **London forces**- are extremely weak forces that occur between any two molecules or atoms and are the result of temporary dipole moments. For example, two nitrogen molecules (N_2, 78% of air) can be briefly attracted to each other by a distortion of the electron cloud resulting in a very weak electrostatic attraction (Fig. **6**).

$$N\equiv N \quad N\equiv N$$
$$\delta^+ \delta^- \quad\quad \delta^+ \delta^-$$

Figure 6: For an instant (*i.e.* $<10^{-10}$ seconds), a temporary dipole in one nitrogen molecule is attracted to a temporary dipole moment in another N_2 molecule. This is referred to as a London Force.

Pre-Lab Exercises

All structures built in this exercise will be transferred to your report. Use arrows to identify key bonds or locations. A special type of covalent bond that will be widely used in this exercise is a peptide bond, which links two amino acids together.

a. In Spartan, build Asp and Gln amino acids separately.

b. Using Semi empirical (PM3) calculations, estimate the dipole moment of each amino acid (Asp and Gln).

c. Now build a simple peptide composed of one Asp and one Gln and, in your report, indicate where the peptide bond is located.

d. Two types of hydrogen bonding exist in peptide sheets (alpha, beta). Include a brief description of the nature of each type.

e.   Carbon has three allotropes (graphite, diamond, fullerenes). Nanotubes are rolled up sheets of carbon that form tubes with small diameters (*i.e.* < 10 nm). Provide the hybridization and a brief (2-3 sentences) description of the structure of diamond, graphite, and a spherical fullerene (*i.e.* C_{60}).

Exercise One

This exercise assumes you have completed previous exercises in Spartan and are familiar with the software package. It also assumes you have some knowledge or familiarity with Excel.

1.   In separate files, build F_2, Cl_2, Br_2 and I_2. When running the structure use Single point energy, Semiempirical (PM3). In your explanation box, answer:

   a.   What type of bond connects the atoms in each halogen?

   b.   Offer an explanation for the change in bond distance.

2.   In separate files, build N_2, N_2H_2, N_2H_4 and remember nitrogen likes 3 bonds (triple, single and a double or three single) and hydrogen likes one bond.

   a.   What type of bond connects the nitrogen atoms?

   b.   Offer an explanation for the shift in N, N bond distance in each structure.

3.   In separate files build the structures LiF, NaF and KF and recall the halides like to form a single bond.

   a.   What type of bond forms the alkali-halide salt?

   b.   Offer an explanation for the differences in bond distances.

4.   Build an acetate molecule (CH_3COO^-, charge = anion). In the same screen, select a Ca atom (from ENT tab) with two single bonds and

connect one to each oxygen atom on the carboxylate (see Fig. **2** above). When setting up the calculations (Semiempirical, PM3) be sure set the charge to +1 for the complex (-1 carboxylate, +2 Calcium). Once the structural calculation is completed, copy the image to your worksheet and measure and record the Ca-O bond distances. Now perform the same calculations for Mg^{+2}-acetate, Sr^{+2}-acetate and Ba^{+2}-acetate, copy the finished structure to your worksheet and measuring the cation-oxygen bond distances.

a.   What type of bond is the alkaline earth cation-acetate bond?

b.   Explain any trends calculated involving the cation-oxygen bond.

5.   Build an H-Br molecule and minimize it. Copy and paste the structure (holding down both the left and right buttons on the mouse) until you have ten H-Br molecules in the same window (note when pasting in Spartan, structures are copied on themselves so several structures may be present but they appear as one). Move the structure so they are close to each other and then minimize the congregation of molecules.

a.   What type of force causes the molecules to be attracted to each other?

b.   Measure and record at least six bond distances involving dipole-dipole interactions involving a –H on one atom and a –Br on another. How does these distances compare to those of covalent bonds (*i.e.* N=N, *etc.*)?

6.   Build a H_2O molecule and minimize it. Copy and paste the structure (holding down both the left and right buttons on the mouse) until you have ten water molecules in the same window (note when pasting in Spartan, structures are copied on themselves so several structures may be present but they appear as one). Move the structure so they are close to each other and then minimize the congregation of molecules.

a.   What type of force causes the molecules to be attracted to each other?

b. Measure and record at least six bond distances involving dipole-dipole interactions involving a –H on one atom and a –O on another. How does these distances compare to those of covalent bonds?

7. Build a N_2 molecule and minimize it. Copy and paste the structure (holding down both the left and right buttons on the mouse) until you have ten nitrogen molecules in the same window (note when pasting in Spartan, structures are copied on themselves so several structures may be present but they appear as one). Move the structure so they are close to each other and then minimize the congregation of molecules.

a. Is there are evidence they are attracted to each other?

b. If so, what type of force causes the molecules to be attracted to each other?

c. Measure and record at least six bond distances involving dipole-dipole interactions involving an N on one atom and a N on another.

Table 1: After the pre-lab exercises are answered, construct a table in WORD that appears as follows (make boxes as large as needed). Copy your Spartan images into the structure column (use a while background) and measure the distance of each (in Angstroms).

Species	Structure	Distance (A)	Explanation
1. F_2, Cl_2, Br_2, I_2			
2. N_2, N_2H_2, N_2H_4			
3. LiF, NaF, KF			
4. Mg-Ac, Ca-Ac, Sr-Ac, Ba-Ac			
5. HBr			
6. H_2O			
7. N_2			

Advanced Exercise: Stretching a peptide using a nanotube

First you will learn to construct a (10, 0) nanotube. You'll find that these set of tubes (10,0; 12,0; 14,0; *etc.*) are pretty straight forward to construct in this molecular modelling software. The instructions are:

1. Click on the ENT Tab and select the carbon that is sp^2 hybridized.

2. Using only sp^2 hybridized carbon atoms, make a ten carbon atom chain. Be sure that all of the double bonds are contained in the chain and you have no protruding double bonds. (Fig. **7A**)

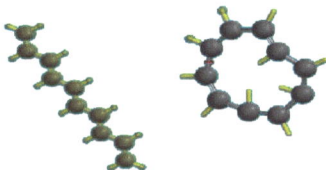

Figure 7: (A) A ten-carbon chain (B) the chain is closed to form a ring which is the basic repeating unit for a (10, 0) nanotube.

3. Now connect the two ends of the chain to form a ring (Fig. **7B**).

4. Copy and paste a ring in the same workspace and connect every other bond forming a portion of a (10, 0) ring. There should be a series of 6-membered rings. (Fig. **8**)

Figure 8: The nanotube geometry begins to take shape.

5. Copy and paste the subunit (2 rings linked) and connect every bond to form a mini-tube with four rings. Copy and paste the four ring units, connect the bonds and form an eight ring unit. Repeat this unit you have a nanotube that is approximately 6-7 nm long (Fig. **9**).

(D)

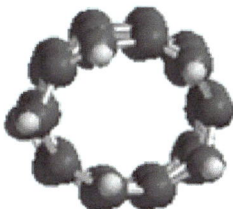

Figure 9: An end on view of the carbon structure illustrated its tubular geometry.

Part B

The nanotube will be used as an inert background or template to construct their peptide. By varying the diameter (*i.e.* 10,0), 16,0), (22,0)) students can vary the geometry to the peptide in a predictable fashion. In order to relate the basic types of bonding outlined above, the peptide will be sued to demonstrate a host of bonds.

    a.   Build a Asp-Gln in a sequence of 16 amino acids total (Asp1,Gln2,Asp3…Asp8, Gln8)) in a workspace separate from the nanotube. Leave your nanotube workspace open.

    b.   Copy and paste this peptide structure two times into the nanotube workspace. Connect the protruding bond at the end of the peptide to a protruding bond on the end of the tube. Connect the other end of the peptide to the other end of the peptide in a straight line (see Fig. **10**).

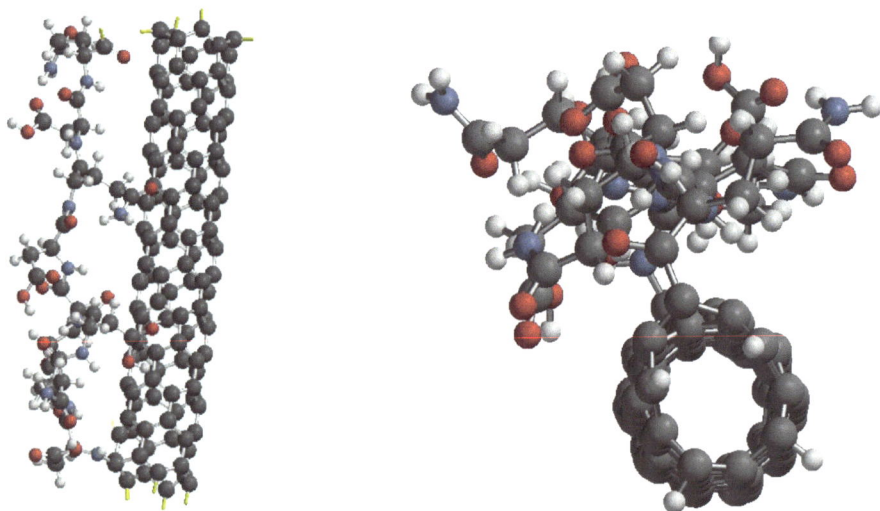

Figure 10: One segment of the peptide is being attached to the nanotube backbone.

    c.   Copy/paste a second peptide to the nanotube workspace and connect the ends of the peptide to the opposite ends of the nanotube in a straight line so you have two peptide chains attached to the nanotube (see Fig. **11**).

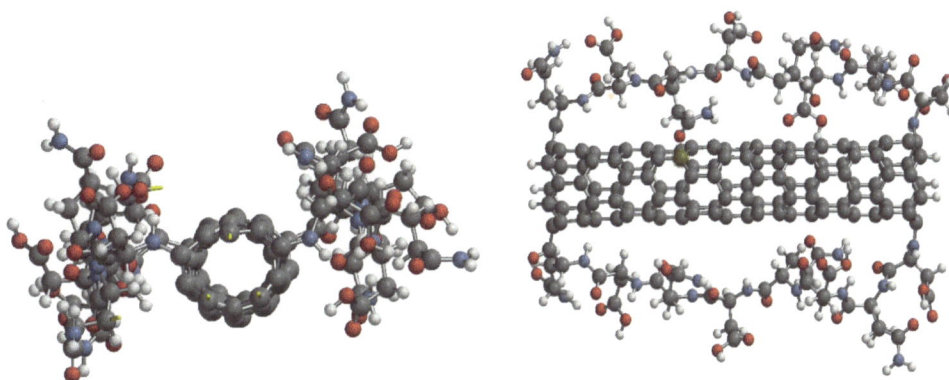

Figure 11: Two peptides are attached to the nanotube template but are not connected to each other – yet!

6. Now copy/paste and connect a third (first) and a fourth (last) peptide to the nanotube backbone. None of the four peptides are connected to each other – yet (Fig. **12**). Save this file as "peptide_nanotube".

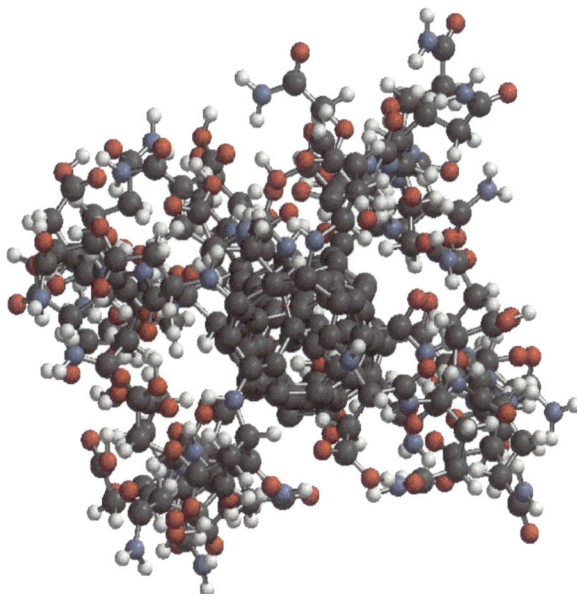

Figure 12: All four peptides are attached to the nanotube by four bonds on either end (8 bonds total). Minimize the energy of this nanotube-peptide structure.

7. Disconnect (break bond command) the eight bonds that are holding the four peptides to the nanotube. Connect the four peptides with three bonds (see Fig. **13**). Remove the nanotube from the newly form peptide and cut it from the workspace so only the new peptide remains. Save this file as "peptide".

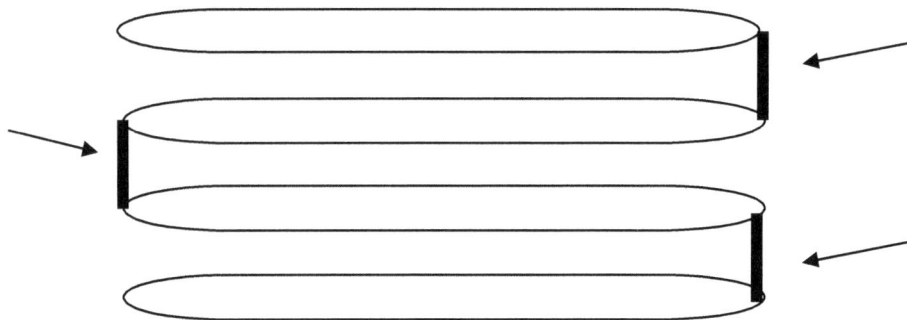

Figure 13: After the peptide is disconnected from the nanotube, three bonds connect the four peptides forming a single larger structure.

8. Construct a Ca(II) ion (semiempircal, PM3, charge=dication). Once the computational work is completed on a single ion, copy and paste it ten times into the peptide structure. Save this structure as "peptide_calcium" and minimize it. Identify which functional group(s) the Ca(II) ions were attracted to and measure the distance between the ion and the group (all ten). Citing Coulombs law, discuss/explain this result in your report.

9. Open the peptide file into a new workspace and save it as peptide-fluoride. In a separate workspace, construct a fluoride (F⁻) ion and perform the typical calculations (recall, charge = anion). Once this calculation is complete, copy/paste ten of the anions into a peptide structure and minimize the systems energy and save the system as peptide_fluoride. Measure and record the ten distances (F- to functional group) and, citing Coulombs law, explain the result.

10. Open the peptide file into a new workspace and save it as peptide-water. In a separate workspace, construct a water molecule and perform the typical calculations (recall, charge = neutral). Once this

calculation is complete, copy/paste ten of the molecules into the peptide structure and minimize the systems energy and save the system as peptide_water. Measure and record the ten distances (water to functional group) and, citing Coulombs law, explain the result. Also, observe and attempt to measure and differences between this structure and the two previous systems (fluoride, calcium). Discuss what impact water had on the structure.

11. Open a fresh peptide workspace. In a second workspace build and run ethanol. Copy and paste ten of the ethanol structures into the peptide workspace and save the system as peptide_ethanol. Minimize the energy and study the interaction between the alcohols. Compare/contract any differences you see between these peptide structures and the structure that resulted when minimized in the presence of water molecules.

12. Be sure to copy/paste all four systems (calcium, fluoride, water, and ethanol) into your report and use arrows, when possible, to help identify key structural changes between the four systems. Parameters such as the molecular volumes, width or length, critical bond angles and general shifts in the peptide structure can be addressed as examples of changes between the different systems.

REFERENCES

[1]   Gillespie, R. J.; Popelier, P. L. A. *Chemical Bonding and Molecular Geometry: From Lewis to Electron Densities*, Oxford University Press - USA Topics in Inorganic Chemistry, **2001**.

[2]   Scanlon, S.; Aggeli, A. Self-assembling peptide nanotubes, *NanoToday*, **2008**, *3 (3 -4)*, 22-30.

[3]   Nikitin, A.; Zhang, Z.; Nilsson, A. Energetics of C–H Bonds Formed at Single-Walled Carbon Nanotubes, *Nano Lett.*, **2009**, *9 (4)*, 1301–1306.

CHAPTER 10

Nuclear Stability Belt

Thomas J. Manning[*] and Aurora P. Gramatges

Department of Chemistry, Valdosta State University, Valdosta, Georgia, USA, and Instituto Superior de Tecnología y Ciencias Aplicadas, La Habana, Cuba

Abstract: In this exercise students will balance nuclear reactions, and will evaluate nuclear stability based on the proton/neutron ratio. Students will construct stability belts using their spreadsheet program.

Keywords: Nuclear chemistry, nuclear reactions, nuclear stability, radioactive decay.

INTRODUCTION

Hands on nuclear chemistry exercises can be difficult to incorporate in an undergraduate course for a number of reasons including safety issues, economics of equipment and supplies, and the potential for unreasonable time scales for most reactions [1]. This exercise will focus on students evaluating nuclear decay data and recreating various aspects of a stability belt [2]. Table **1** provides a list of elements and their stable isotopes [2,3]. Stability tables include only stable (nonradioactive) isotopes. Start your report by answering the questions below.

Pre-Lab Questions

1. Give a balanced nuclear chemical reaction that goes under Beta decay? In terms of proton/neutron ratios, what types of nuclei undergo beta decay to from more stable nuclei?

2. Show the process of electron capture (EC) in a balanced nuclear chemical reaction? How does EC affect the mass number or the atomic number of the element that is going under the decay?

---

*Address correspondence to **Thomas J. Manning**: Department of Chemistry, Valdosta State University Valdosta GA 31698, USA; Tel: 229-333-7178; E-mail: tmanning@valdosta.edu

3. What is a positron? Give a balanced nuclear reaction that demonstrates positron emission? If carbon (^{12}C) will go under positron emission, provide the balanced reaction.

4. What is an alpha particle? How will a U-235 atom undergo alpha emission (provide the balanced nuclear reaction)?

Part I

1. For elements $Z^{\#}$ =1-30 (hydrogen to zinc), plot the number of protons (x-axis) *vs.* the number of neutrons (y-axis) for EACH isotope. Some elements may have more than one isotope (plot p, n values all isotopes). Table **1** gives the raw data needed. Use an x, y scatter plot and fit your data with a linear fit and provide the equation and correlation coefficient on the graph. Remember to label the axis and put your name on the top of the graph before moving to your report. Include a figure caption and comment on the relationship between protons and neutrons for stable nuclei with less than 31 protons. This graph should be at least 12x12 cm in size.

2. On your graph, use an arrow to show where ^{14}C is located and, in a bullet below the figure caption, provide the decay reaction.

3. Iron has some stable isotopes (Fe-54, Fe-56, Fe-57, Fe-58) and some unstable isotopes (Fe-52, Fe-55, Fe-59, Fe-60). Identify where each unstable isotope would fit on the stability belt (use arrow, ➔ Fe-60).And below the graph indicate the decay reaction and its half-life in a bullet below the figure caption.

4. Sodium has one stable isotope (Na-23) and two unstable isotopes (Na-22 and Na-24). Using an arrow, show where each unstable isotope would fall on the graph and below graph give the nuclear decay scheme for each isotope returning to being a stable nuclei.

5. One of the elements on your graph has two isotopes that are given the symbols "D" and "T". In terms of neutron and protons ratios, T is

unique compared to the other elements on the periodic table. What is this uniqueness? What is its nuclear decay reaction?

Part II

The students will now generate a new graph following a similar format to that used in Part I. Use isotopes from the table that are between 31 and 82 protons – include all isotopes. The graph shown below is just an example of protons *vs.* neutrons of elements from Z# 31-82. Again include your y = mx+b equation (linear fit) and its correlation coefficient (and your name) on your graph. Also, your graph should have a figure caption. Fig. **1** shows the general form your stability plots should follow.

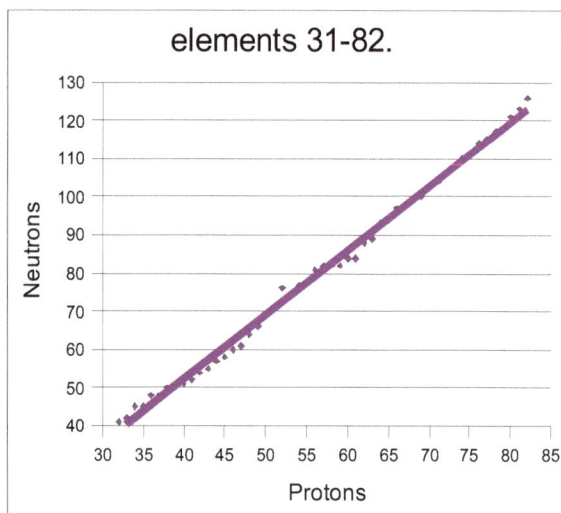

Figure 1: The relationship between protons and neutrons for stable nuclei/elements for elements 31-82.

Consider the slope of the second graph (Z# 31-82) and first data set (Z# 1-30), and answer the following questions.

1. Why are slope values of the two graphs different?

2. Why is slope NOT equal to one (m=1) in the second data set.

3. How many nuclei on the second graph have a proton to neutron ratio of 1.00? If there are any nuclei, which ones are they?

4. Both Sr-90 and Cs-137 are radioactive isotopes that are produced in industrial nuclear reactions. Below your graph, indicate what reactions they come from and the use of the reactions. Using an arrow, indicate where each of these would fall on your stability belt.

5. I-131 is used in nuclear medicine. Describe two applications and also provide its decay reaction and half-life. Indicate the location of this isotope (I-131➔) on your graph.

6. Tc and Pm have no stable isotopes and are not found in the earth's crust. Tc-99m is used in nuclear medicine. Describe its medicinal use, provide the mechanism of its decay reaction, indicate what "m" means in 99m, and show its location on your stability belt.

7. In the region to the left of the stability belt (high n/p ratio) what is the decay mechanism to return to the stability belt (give a sample reaction).

Table 1: An alphabetical list of elements with their stable isotopes. Stability belts are composed of stable nuclei. Use a periodic table to obtain the Z# (protons) for each element.

Name	E.I	mass	E.I	mass	E.I	Mass	E.I	mass	E.I	mass	E.I	mass	E.I	mass
Aluminum	27	27												
Antimony	121	121	Sb-123	123										
Argon	36	36	Ar-38	38	Ar-40	40								
Arsenic	75	74.9												
Barium	130	130	Ba-132	132	Ba-134	134	Ba-135	134.9	Ba-136	136	Ba-137	137	Ba-138	138
Beryllium	9	9.01												
Bismuth	209	209												
Boron	10	10	B-11	11										
Bromine	79	78.9	Br-81	80.9										
Cadmium	106	106	Cd-110	110	Cd-111	111	Cd-112	111.9	Cd-113	113	Cd-114	114	Cd-116	116
Calcium	40	40	Ca-42	42	Ca-43	43	Ca-44	43.96	Ca-46	46	Ca-48	48		
Carbon	12	12	C-13	13										
Cerium	136	136	Ce-138	138	Ce-140	140	Ce-142	141.9						
Cesium	133	133												
Chlorine	35	35	Cl-37	37										
Chromium	50	49.9	Cr-52	51.9	Cr-53	52.9	Cr-54	53.94						
Cobalt	59	58.9												

Table 1: contd...

Copper	63	62.9	Cu-65	64.9										
Dysprosium	156	156	Dy-158	158	Dy-160	160	Dy-161	160.9	Dy-162	162	Dy-163	163	Dy-164	164
Erbium	162	162	Er-164	164	Er-166	166	Er-167	166.9	Er-168	168	Er-170	170		
Europium	151	151	Eu-153	153										
Fluorine	19	19												
Gallium	69	68.9	Ga-71	70.9										
Gadolinium	152	152	Gd-154	154	Gd-155	155	Gd-156	155.9	Gd-157	157	Gd-158	158	Gd-160	160
Germanium	70	69.9	Ge-72	71.9	Ge-73	72.9	Ge-74	73.92	Ge-76	75.9				
Gold	197	197												
Hafnium	174	174	Hf-176	176	Hf-177	177	Hf-178	177.9	Hf-179	179	Hf-180	180		
Helium	3	3.02	He-4	4										
Holmium	165	165												
Hydrogen	1	1.01	H-2	2.01										
Indium	113	113	In115	115										
Iodine	127	127												
Iridium	191	191	Ir-193	193										
Iron	54	53.9	Fe-56	55.9	Fe-57	56.9	Fe-58	57.93						
Krypton	78	77.9	Kr-80	79.9	Kr-82	81.9	Kr-83	82.91	Kr-84	83.9	Kr-86	85.9		
Lanthanum	138	138	La-139	139										
Lead	204	204	Pb-206	206	Pb-207	207	Pb-208	208						
Lithium	6	6.02	Li-7	0.02										
Lutetium	175	175	Lu-176	176										
Magnesium	24	24	Mg-25	25	Mg-26	26								
Manganese	55	54.9												
Mercury	196	196	Hg-198	198	Hg-199	199	Hg-200	200	Hg-201	201	Hg-202	202	Hg-204	204
Molybdenum	92	91.9	Mo-94	93.9	Mo-95	94.9	Mo-96	95.9	Mo-97	96.9	Mo-98	97.9	Mo-100	99.9
Neodymium	142	142	Nd-143	143	Nd-144	144	Nd-145	144.9	Nd-146	146	Nd-148	148	Nd-150	150
Neon	20	20	Ne-21	21	Ne-22	22								
Nickel	58	57.9	Ni-60	59.9	Ni-61	60.9	Ni-62	61.93	Ni-64	63.9				
Niobium	93	92.9												
Nitrogen	14	14	N-15	15										
Osmium	184	184	Os-186	186	Os-187	187	Os-188	188	Os-189	189	Os-190	190	Os-192	192
Oxygen	16	16	O-17	17	O-18	18								
Palladium	102	102	Pd-104	104	Pd-105	105	Pd-106	105.9	Pd-108	108	Pd-110	110		
Phosphorous	31	31												
Platinum	190	190	Pt-192	192	Pt-194	194	Pt-195	195	Pt-196	196	Pt-198	198		
Potassium	39	39	K-40	40	K-41	41								
Praseodymium	141	141												
Rhenium	185	185	Re-187	187										

Table 1: contd...

Element														
Rhodium	103	103												
Rubidium	85	84.9	Rb-87	86.9										
Ruthenium	96	95.9	Ru-98	97.9	Ru-99	98.9	Ru-100	99.9	Ru-101	101	Ru-102	102	Ru-104	104
Samarium	144	144	Sm-147	147	Sm-148	148	Sm-149	148.9	Sm-150	150	Sm-152	152	Sm-154	154
Scandium	45	45												
Selenium	74	73.9	Se-76	75.9	Se-77	76.9	Se-78	77.92	Se-80	79.9	Se-82	81.9		
Silicon	28	28	Si-29	29	Si-30	30								
Silver	107	107	Ag-109	109										
Sodium	23	23												
Strontium	84	83.9	Sr-86	85.9	Sr-87	86.9	Sr-88	87.91						
Sulfur	32	32	S-33	33	S-34	34	S-36	35.97						
Tantalum	180	180	Ta-181	181										
Tellurium	122	122	Te-123	123	Te-124	124	Te-125	124.9	Te-126	126	Te-128	128	Te-130	130
Terbium	159	159												
Thallium	203	203	Tl-205	205										
Thorium	232	232												
Thulium	169	169												
Tin	112	112	Sn-114	114	Sn-115	115	Sn-116	115.9	Sn-117	117	Sn-118	118	Sn-119	119
Titanium	46	46	Ti-47	47	Ti-48	47.9	Ti-49	48.95	Ti-50	49.9				
Tungsten	180	180	W-182	182	W-183	183	W-184	184	W-186	186				
Uranium	234	234	U-235	235	U-238	238								
Vanadium	50	49.9	V-51	50.9										
Xenon	124	124	Xe-126	126	Xe-128	128	Xe-129	128.9	Xe-130	130	Xe-131	131	Xe-132	132
Ytterbium	168	168	Yb-170	170	Yb-171	171	Yb-172	171.9	Yb-173	173	Yb-174	174	Yb-176	176
Yttrium	89	88.9												
Zinc	64	63.9	Zn-66	65.9	Zn-67	66.9	Zn-68	67.92	Zn-70	69.9				
Zirconium	90	89.9	Zr-91	90.9	Zr-92	91.9	Zr-94	93.91	Zr-96	95.9				

*E-I stands for Element isotope.

* This is not a full table of isotopes

Part III

If you use google.com and enter a specific isotope (*i.e.* uranium-235), you can find needed data easily.

a. Table **2** lists some elements and isotopes that have more than 83 protons. Note that Z>83 are not be listed on stability belts. Why? (Hint, with Z>83, what type of decay do they undergo?). For radium, radon, thorium, uranium and plutonium, find the isotopes for each of the isotopes listed (all will be unstable) and develop a table that follows the format shown in Table **2**.

b. I-131, Am-245, Co-60, Cs-137, Sr-90 are radioactive isotopes that either have industrial/medical applications or have been of great concern in environmental pollution. Briefly describe the role of each in its major (best known) activity.

Table 2: In your report develop a table that follows this format. It is important to know the mechanism for which Z# > 83 return to the stability belt.

Element/ isotope	# Protons	# neutrons	n/p ratio	Decay reaction	Half-life (include units on time)
U-235	92	146		U-238 ➔ Th-234 + He-4	
U-238					
U-242					
Np-225					
Np-229					
Pu-232					
Pu- 228					
Am- 235					
Am-238					
Pa-215					

REFERENCES

[1]   Katz, S. A.; Bryan, J. C. *Experiments in Nuclear Science* CRC Press, Taylor and Francis Group, **2011**.

[2]   Loveland, W. D.; Morrissey, D. J.; Seaborg, G. T. *Modern Nuclear Chemistry,* John Wiley & Sons, Inc. Hooboken, New Jersey, **2006.**

[3]   Friedlander, G.; Kennedy, J. W.; Macias, E. S.; Miller, J. M. *Nuclear and Radiochemistry* 3rd ed., John Wiley & Sons, Inc, **1981**.

CHAPTER 11

Speciation Plots and pH

Thomas J. Manning* and Aurora P. Gramatges

Department of Chemistry, Valdosta State University, Valdosta, Georgia, USA, and Instituto Superior de Tecnología y Ciencias Aplicadas, La Habana, Cuba

Abstract: In this exercise students will review fundamental aspects of acid/base chemistry in the aqueous phase. Students will also simulate the impact that shifting or altering pH has on a monoprotic or diprotic or polyprotic species.

Keywords: Acid/base chemistry, speciation, equilibrium constant, pH.

INTRODUCTION

The acidity or basicity of a system can significantly impact or alter the chemistry that takes place in a beaker, a living organism or an ecosystem [1-3]. For example, iron metal will dissolve in a very acidic medium and form iron(II) or iron(III) that will exists in the aqueous phase. If the pH is shifted to being more basic, than the iron will precipitate out as an oxide (*i.e.* FeO), hydroxide ($Fe(OH)_2$), or an oxyhydroxide (FeOOH). In this exercise, the student will look at four acids and plot the species present from a pH=0 to a pH=14. First, the student will follow step-by-step instructions in a spreadsheet to simulate the deprotonation of acetic acid (HAc) to form acetate (Ac^-) as a function of pH. The commands given here are for Excel.

$$HAc(aq) + H_2O(l) \rightarrow H_3O^+(aq) + Ac^-(aq)$$

And the equilibrium expression is:

$$K_a = \frac{[H_3O^+][Ac^-]}{[HAc]} = 1.8 \times 10^{-5}$$

This can be expanded into the Henderson-Hasselbalch (H-H) equation, which is typically used for buffers.

---

*Address correspondence to Thomas J. Manning: Department of Chemistry, Valdosta State University Valdosta GA 31698, USA; Tel: 229-333-7178; E-mail: tmanning@valdosta.edu

$$pH = pK_a + \log_{10}[Ac^-]/[HAc]$$

In this simulation, the pH and pK_a are defined at each point. The value of the pH will increase from 0 to 14 in increments of 0.1 (0, 0.1, 0.2, *etc.*) while the pK_a (4.74) is the same for all points. This allows us to rearrange the H-H equation:

$$10^{(pH-pKa)} = [Ac^-]/[HAc]$$

This equation can be redefined by

$$10^{(pH-pKa)} = [X]/[C-X]$$

where C is the starting concentration of HAc and X is the amount of HAc that is deprotonated and forms acetate. This equation can be rearranged:

$$\frac{C*10^{(pH-pKa)}}{1+10^{(pH-pKa)}} = X$$

You will now use this equation to simulate a speciation plot for the acetic acid, acetate species.

Pre-Lab questions

First answer the pre-lab questions in your report, followed by copies of your graphs (with a figure caption). When done with the entire exercise you should have four graphs (two monoprotic acids, two diprotic acids).

1. Provide the name and empirical formula for the six common strong acids.

2. Provide the name and empirical formula for six common strong bases.

3. Do strong acids and strong bases have equilibrium constants? Explain.

4. For the following monoprotic weak acids, write the reaction and equilibrium equation and include their K_a values (see equations above for the form).

    a.   Hydrofluoric acid

    b.   Nitrous acid

    c.   Hydrocyanic acid

    d.   Ammonium

    e.   Formic acid

5. For the following diprotic or triprotic weak acids, write the reaction for each deprotonation and the corresponding equilibrium equation and include the K_a value for each proton (see equations above for the form).

    a.   Carbonic acid

    b.   Sulfurous acid

    c.   Oxalic acid

    d.   Phosphoric acid

Part 1 and 2 are monoprotic acids and part 3 and part 4 are diprotic acids. Your instructor will give directions for which species to plot. The instructions are provided for the first monoprotic acid and the first diprotic acid. Use the instructions/format provided for HAc and H_2SO_3 for the second set of acids.

Part 1

1. Open a new spreadsheet.

2. In location A1 place the header "pH"

3. In location A2 enter the value 0.

4. In location A3 enter the equation "=sum(A2+0.1)"

5. Copy and paste this equation down to A142. Your last value (A142) should be 14.

6. In location B1 enter the header "pKa"

7. In location B2 enter the value 4.74 and copy this value down to B142. The same value should appear in all locations.

8. In C1 enter the header "pH - pKa"

9. In location C2 enter the equation "=SUM(A2-B2)" and copy it down to C142. In C2 you should have the value -4.74 and in the last location (C142) you should have the value 9.26 (using $pK_a + pK_b = 14$, what is the pK_b of acetic acid/acetate?).

10. In location D1 enter the header "pH/pKa; exp". The equation that will be entered in this column will be part of equation.

11. In location D2 enter the equation "=EXP(C2)" and copy it down to D142.

12. In location E1 enter the header "Init. Conc. Acetic" which stands for the initial concentration of acetic acid. Be sure to make you columns wide enough to clearly read the header.

13. In location E2 enter the number "1" and copy it down to E142. You are starting with 1 M acetic acid. A small fraction of this will be dissociated in an acidic pH but as the solutions acidity decreases and its basicity increases the fraction of acetic acid will decrease and the amount of acetate will increase.

14. In location F1 enter the header "Acetate Conc".

15. In location F2 enter the equation "=SUM(E2*D2/(1+D2))" and copy/paste it down to F142. This is equation 6 from above.

16. In location G1 enter the header "Acetic Acid Eq. Conc". This column will calculate the equilibrium concentration of acetic acid (HAc) at each pH or $[H_3O^+]$.

17. In location G2 enter the equation "=SUM(E2-F2)" and copy it down to G142. Your first value (G2) should be approximately 0.991 and your last value (G142) should be approximately 0.00009.

18. You will now create a graph with two series using the chart wizard. In the first series make the x-axis the pH values (A2…A142) and the y-axis the acetate concentration values (F2…F142). In the second series use the same pH values (A2…A142) for the x-axis and use the acetic acid concentration values (G2…G142) for the y-axis.

19. Label the x-axis (pH) and y-axis (Concentration) and use your name for the title on top of the graph. Be sure the pH axis is labeled every unit from 0-14. Your graph should look like that shown in Fig. **1**.

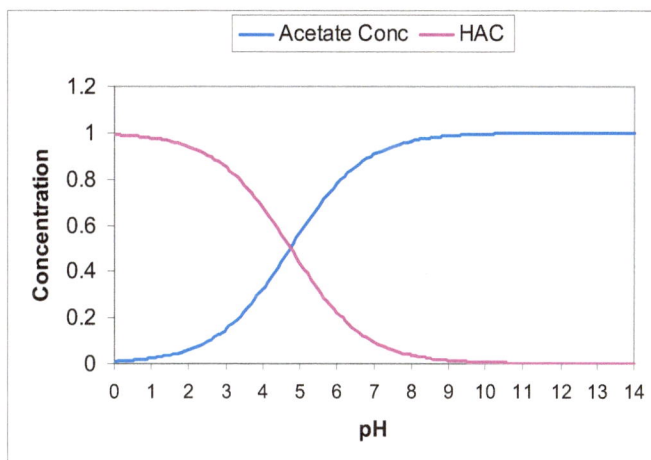

Figure 1: Speciation plot generated in a spreadsheet for the acetic acid and acetate pair.

Part II

Construct a speciation plot for hydrofluoric acid (HF) using the same directions as utilized in **Part 1**. Use a concentration of 0.5 M and calculate the pK_a from the K_a above.

Part III

This exercise will focus on generating a speciation plot for the diprotic acid H_2SO_3.

1. Open a new Excel spreadsheet.

2. In location A1 place the header "pH"

3. In location A2 enter the value 0.

4. In location A3 enter the equation "=sum(A2+0.1)"

5. Copy and paste this equation down to A142. Your last value (A142) should be 14.

6. In location B1 enter the header "pKa"

7. In location B2 enter the value 1.85 and copy this value down to B142. The same value should appear in all locations.

8. In C1 enter the header "pH - pKa"

9. In location C2 enter the equation "=SUM(A2-B2)" and copy it down to C142. In C2 you should have the value -1.85 and in the last location (C142) you should have the value 12.15.

10. In location D1 enter the header "pH/pKa; exp". The equation that will be entered in this column will be part of equation.

11. In location D2 enter the equation "=EXP(C2)" and copy it down to D142.

12. In location E1 enter the header "Init. Conc. Sulfurous" which stands for the initial concentration of sulfurous acid. Be sure to make you columns wide enough to clearly read the header.

13. In location E2 enter the number ".5" and copy it down to E142. You are starting with 0.5 M. A small fraction of this will be dissociated in an acidic pH but as the solutions acidity decreases and its basicity increases the fraction of acetic acid will decrease and the amount of acetate will increase.

14. In location F1 enter the header "HSO3- Conc".

15. In location F2 enter the equation "=SUM(E2*D2/(1+D2))" and copy/paste it down to F142.

16. In location G1 enter the header "Sulf. Acid Eq. Conc". This column will calculate the equilibrium concentration of sulfurous acid (H_2SO_3) at each pH or [H_3O^+].

17. In location G2 enter the equation "=SUM(E2-F2)" and copy it down to G142. Your first value (G2) should be approximately 0.43 and your last value (G142) should be approximately 0.0000026.

18. For the sake of clarity, skip column H. Because sulfurous acid has two pK_a's there will be three species (H_2SO_3, HSO_3^-, SO_3^{-2}) to represent.

19. In location I1 enter the header "pKa2".

20. In location I2 enter the value 7.20 and copy it to I142.

21. In location J1 enter the header "pH-pKa2".

22. In location J2 enter the equation "=SUM(A2-I2)" and copy it down to J142.

23. In location K1 enter the header "pH/pK2, exp".

24. In location K2 enter the formula "=EXP(J2)" and copy it down to K142

25. In L1 enter the header "Initial SO3-2" which represents the initial sulfite concentration.

26. In location L2 enter the equation "=SUM(K2*F2/(K2+1))". What equation in the lab introduction does this represent? What variables does K2 and F2 represent? Copy this equation down to L142. The last value (L142) should be about 0.499.

27. In location M1 enter the header "Final HSO_3" which represents the final HSO_3^- concentration at each pH value.

28. In location M2 enter the equation "=SUM(E2-G2-L2)" and copy it down to M142. What does the variable E2, G2 and L2 represent?

29. You will now create a graph with the chart wizard which will have three series (SO_3^{-2}, HSO_3^-, H_2SO_3) using the chart wizard. All three series will use the pH values (A2…A142) for the x-axis. H_2SO_3 will be plotted using G2.G142 for the y-axis, HSO_3^- will be plotted using the data in M2.M142, and SO_3^{-2} will be plotted using the data in locations L2…L142.

30. Label the x-axis (pH) and y-axis (Concentration) and use your name for the title on top of the graph. Be sure the pH axis is labeled every unit from 0-14 (*i.e.* 0,1,2,3.). Your graph should look like that shown in Fig. **2** (except with YOUR name on top!).

31. Be sure to copy and paste this graph into your report.

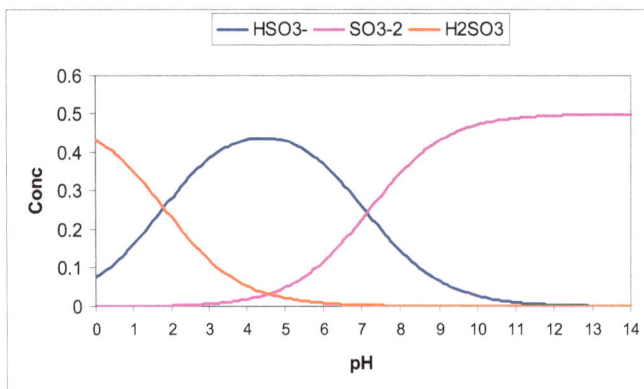

Figure 2: The speciation of sulfite over the pH range 0-14. The starting concentration for H_2SO_3 is 0.15 M.

Part IV

Construct a speciation plot for oxalic acid using the same format/directions utilized in **Part III**. Use a concentration of 0.75 M and calculate the pK_a's from the K_a in the pre-lab. Be sure to copy and paste this graph into your report.

Post Lab Questions

Type the answers to these questions into your report.

a. Why did you always use molar concentrations when adding and subtracting quantities and not use moles? Specifically, you were never provided with a volume for any of the speciation plots. Explain why the volume and subsequently moles (moles = MV) are not needed.

b. EDTA can be a hexaprotic acid. Draw the structure for the hexaprotic acid (assume both nitrogen's are protonated). What does EDTA stand for? List the pK_a's you can find for EDTA (you might not be able to find all 6!).

c. EDTA is known as an aminocarboxylate. DTPA is another well-known aminocarboxylate. Draw its structure and label the eight sites that can protonate. Gd(III)-DTPA (Gd is the lanthanide Gadolinium) is used in MRI or magnetic resonance imaging (MRI), a medical technique. What is Gd(III)-DTPA used for in MRI?

d. In equations in the introduction, the equilibrium expression was quickly shown to be related to the Henderson-Hasselbalch equation. In a step-by-step fashion, show the derivation from one equation to the other equation.

REFERENCES

[1]   Turner, D.R.; Whitfield, M.; Dickson, A.G., The equilibrium speciation of dissolved components in freshwater and sea water at 25°C and 1 atm pressure, *Geochim. Cosmochim. Ac.* **1981**, *45*, 855–881.
[2]   Box, K. J.; Donkor, R. E.; Jupp, P. A.; Leader, I. P.; Trew, D. F.; Turner, C. H., The chemistry of multi-protic drugs: Part 1: A potentiometric, multi-wavelength UV and NMR pH titrimetric study of the micro-speciation of SKI-606, *J. Pharmaceut. Biomed.*, **2008**, *47*, 303-311.
[3]   Hanrahan, G. *Key Concepts in Environmental Chemistry, Chapter 3 – Aqueous Chemistry*, **2012**.

Send Orders of Reprints at reprints@benthamscience.net
Computer Based Projects for a Chemistry Curriculum, 2013, 106-118

CHAPTER 12

The Single Molecule Magnets Mn_{12} and Fe_8

Thomas J. Manning[*] and Aurora P. Gramatges

Department of Chemistry, Valdosta State University, Valdosta, Georgia, USA, and Instituto Superior de Tecnología y Ciencias Aplicadas, La Habana, Cuba

Abstract: In this exercise students will review a number of fundamental concepts including molecular geometry, electron configurations, magnetism, metal-ligand interactions, and material science. Students will use molecular modelling software to build, visualize and study a cutting edge material (single molecule magnet).

Keywords: Molecular geometry, magnetism, molecular modeling, molecular magnets, nanomaterials.

INTRODUCTION

In this exercise molecular modelling software is used to construct two complex structures, Mn_{12} and Fe_8, which are also known as single molecule magnets [1]. In order to be a single-molecule magnet, the object must exhibit a net magnetic spin and have negligible magnetic interactions between its molecules. These single-molecule magnets are being widely investigated in nanomaterial research [2,3]. Scientists believe single molecule magnets have promise in the realization of the smallest practical unit capable of magnetic memory. This is due to their typically large bi-stable spin anisotropy. Additionally, these molecular magnets have given scientists a useful material to study various aspects of quantum mechanics.

This interdisciplinary exercise incorporates a number of topics touched on in general chemistry including magnetism, molecular geometries, hybridizations, and material science, nanotechnology, and oxidation states. In your report, answer the pre-lab questions at the beginning of your report and use a 2D art program for drawings. After constructing Fe_8 and Mn_{12} in Spartan, include at least three

---

*Address correspondence to **Thomas J. Manning**: Department of Chemistry, Valdosta State University Valdosta GA 31698, USA; Tel: 229-333-7178; E-mail: tmanning@valdosta.edu

different images of each structure (from different angles) and measure the required angles and bond distances (see Tables **1**, **2**).

Pre-Lab Questions

On the first page of your report, answer to the following questions.

1. Provide the electron configurations of Mn, Mn^{+3}, and Mn^{+4}

2. How many unpaired electrons are in each atom Mn^{+3} and Mn^{+4}?

3. With four Mn^{+4} ions and eight Mn^{+3} ions, how many unpaired

4. electrons can Mn_{12} potentially have at one time?

5. Define diamagnetic, paramagnetic, and ferromagnetic.

6. Is Mn_{12} diamagnetic? paramagnetic? Or ferromagnetic? Why?

7. Define what constitutes a Single Molecule Magnet? A Quantum Computer?

8. Fe_8 is the abbreviation for Iron(8+), dodeca-hydroxyhexakis(octahydro-1H-1,4,7-triazonine-N1,N4,N7)di-3-oxoocta-, octabromide, nonahydrate. Identify three smaller species (molecules) found within the structure.

9. Mn_{12} is the abbreviation for $Mn_{12}O_{12}(CH_3COO)_{16}(H_2O)_4]2CH_3COOH.4H_2O$. Identify four small ionic or molecular species present in the molecule.

10. Define a coordination number? A ligand? A monodentate ligand? How does octahedral geometry appear (Draw in 2D)?

Mn_{12}-acetate is composed of 4 waters, 16 acetate molecules, 12 Mn atoms (III & IV) with octahedral geometries and 12 oxides ions. Mn_{12} has the empirical formula $Mn_{12}O_{12}(CH_3COO)_{16}(H_2O)_4$. This molecule contains four $Mn^{+4}(S=3/2)$ ions in a central tetrahedron surrounded by eight $Mn^{+3}(S=2)$ ions. S is the spin number and is related to electron spin. Mn_{12} contains oxygen bridges that allow super-exchange coupling among the Mn ions.

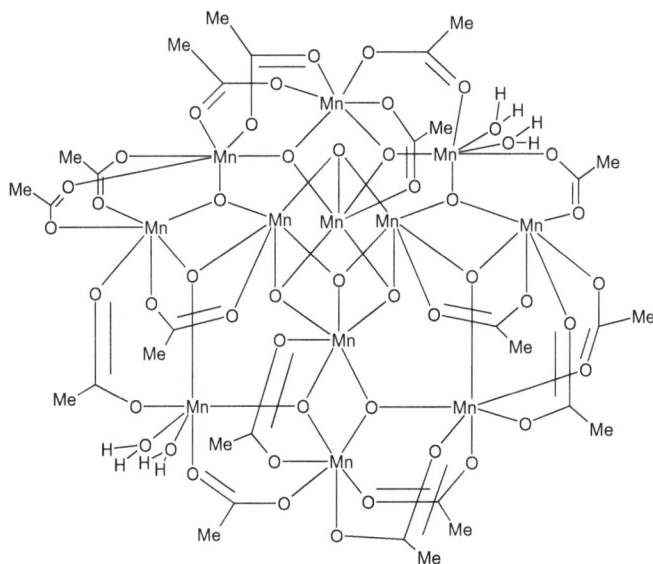

Figure 1: A 2-D image of the Single Molecule Magnet Mn_{12}.

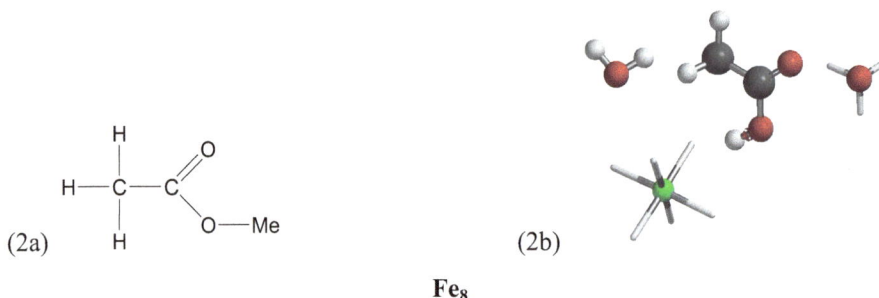

(2a)   (2b)

Fe$_8$

Figure 2a: Acetate (CH_3COO^-) is a key ligand. The carboxylate binds to the positively charged Mn cations. The "Me" is a methyl group ($-CH_3$). **(2b)** (below) The Mn_{12}-cluster is composed of water, acetate, Mn ions, and oxides. The oxide used in building Mn_{12} has 3 bonds, the 3rd being an electrostatic attraction (typically oxygen has 2 bonds).

To prepare the student to build Mn_{12}, they will first build the smaller single molecule magnet Fe_8 in a step-by-step fashion. This molecule will be constructed in Spartan and copied to their reports. This molecule will be constructed in modular sections and then assemble them to form a complete structure. Students in general chemistry may not understand line or stick representations for organic structures. Fig. **3** illustrates a common structural abbreviation system used in Fig. **4**.

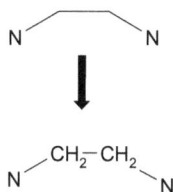

Figure 3: In organic chemistry, bends in straight lines (top) represent carbon atoms with hydrogen atoms attached (bottom). This ethylene structure is found in Fe$_8$.

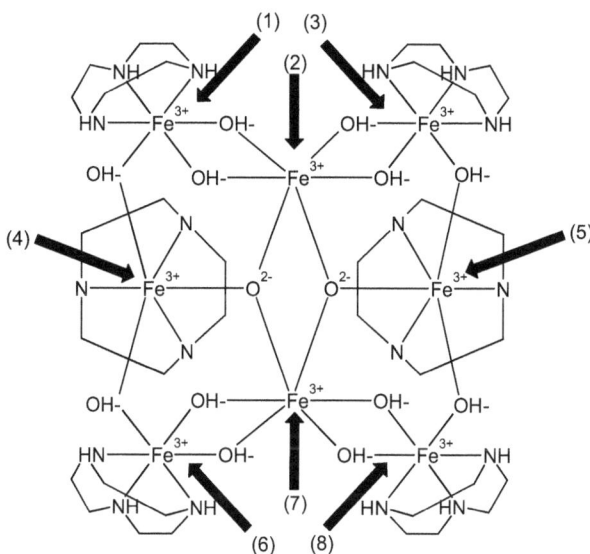

Figure 4: The Single Molecule Magnet Fe$_8$ has the empirical formula C$_{36}$H$_{102}$Fe$_8$N$_{18}$O$_{14}$ 8Br9H$_2$O. The Iron atoms have been pre-labeled 1-8 to aide construction. When you build mn12, the Mn atoms should also be numbered in a sequential fashion.

Exercise I. Building Fe$_8$

In these instructions it is assumed students have already completed other Spartan exercises and are familiar with various molecular geometries (octahedral, trigonal planar, *etc.*). The Fe atoms within the structure have been pre-labeled for the student in Fig. **4**. While it is possible to designate an oxidation for a specific atom in Spartan before any calculations are performed, this exercise is focused on basic geometric factors so oxidation states are not needed at this point. The instructions for constructing Fe$_8$ are:

1) Open Spartan and click on "File", and choose "New" to start a new molecule.

2) Fe_8 will be assembled in sections. Construction will start in the upper left quadrant of the molecule (Fig. **4**) centered on the Fe atom designated #1. Select the Fe atom from the "Exp". tab. Give the Fe atom a six bond configuration by clicking on the octahedral geometry, all with single bonds. Once inserted in the workspace it should appear as Fig. **5**.

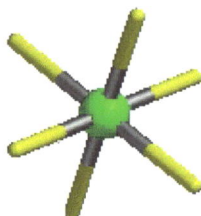

Figure 5: Iron has an octahedral geometry.

3) Nitrogen and oxygen atoms are linked to the Fe atom. Select a sp^3 (tetrahedral geometry) nitrogen listed under the "Exp". tab. Place three N atoms on the Fe atom #1 (Fig. **6**). You can change the color on the specific elements by clicking on the element of choice and entering "options" and "color".

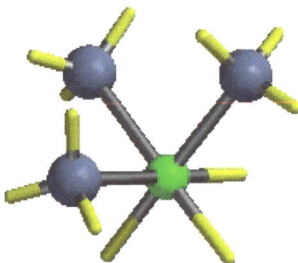

Figure 6: Fe (green) is bound to three Nitrogen atoms (blue).

4) Attach three oxygen atoms to the remaining binding sites on Fe. When selecting the oxygen geometry, select trigonal planar with single bonds (Fig. **7**).

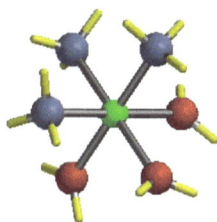

Figure 7: Oxygen atoms (red) are bound to the central iron atom.

5) Two sp^3 hybridized carbon atoms will bridge the nitrogen atoms.
    Connect two carbon atoms (two separate methyl groups, $-CH_3$) to each
    nitrogen atom on your Fe #1 structure, leaving one bonding site on
    each nitrogen atom open (see Fig. **8**).

Figure 8: Two methyl groups (6 total) are attached to each nitrogen atom.

6) Using the above image as a guide, number the carbon atoms one
    through six, starting in the lower left corner and going clockwise
    around the image. Connect C1 to C6, C2 to C3, and C4 to C5,
    resulting in the structure shown in Fig. **9**. Save your structure after the
    addition of every 2 or 3 atoms.

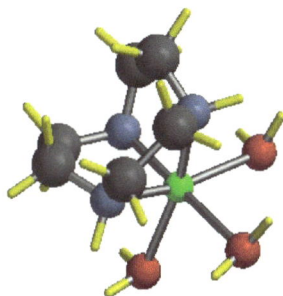

Figure 9: Carbon atoms on different nitrogen atoms are connected.

7) Click "File" and "Save as" to save your molecule. Save the molecule as "Fragment".

8) Paying close attention to the 2D Fe_8 structure (Fig. **4**), note the subcomponent shown in Fig. **9** is repeated four times. Fig. **9** represents the cluster centered on Fe #1, as well as Fe #3, Fe #6, and Fe #8. Copy and paste your structure and rotate it to the upper right corner. They should appear like that shown in Fig. **10**.

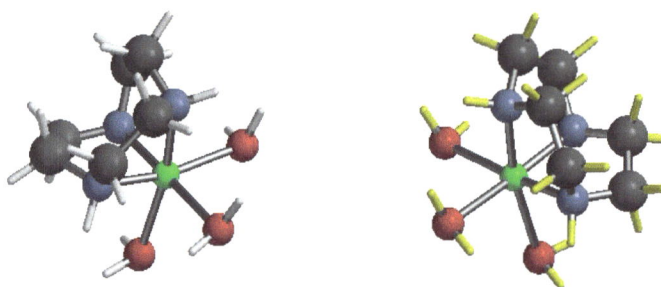

Figure 10: The structure shown in Fig. **9** is copied/pasted and rotated into this position.

9) Referring to Fig. **2**, add Fe atom #2 (octahedral) so it bonds to two oxygen atoms on each cluster (Fig. **11**), serving as a bridge at this point. An oxygen atom on each cluster should be facing in the same direction as the free bonds on the central Fe atom (#2).

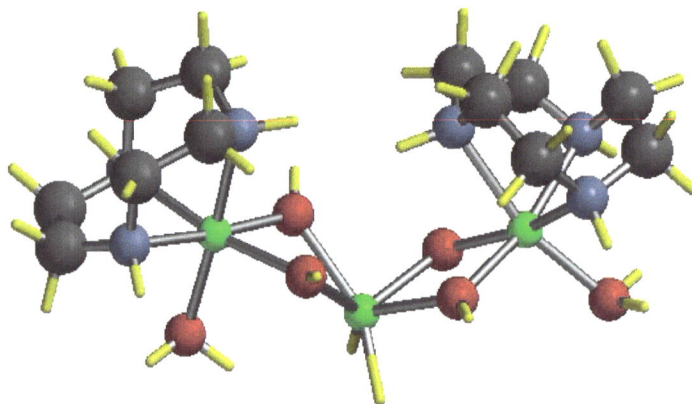

Figure 11: The central iron has two Fe-O bonds to each cluster. It two remaining electron pairs or bonding areas are facing in the same direction as the oxygen atoms.

10. Copy and paste your complex (Fig. **12**) into the workspace and rotate the structure 180° using the same method as in step 13.

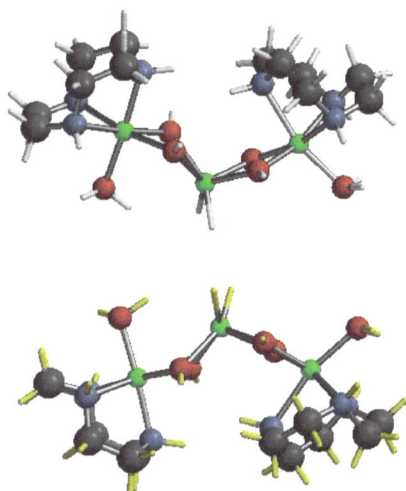

Figure 12: Copy/rotate and rotate your structure until it appears like the above species.

11. Bridge the two complexes together using oxygen atoms that are in trigonal planar (3 bonds) geometry. The two Fe atoms involved in this new bridge are Fe #2 and Fe #7. DO NOT MINIMIZE after this step but save this structure naming it "Halves-bonded". Your structure should appear as it does in Fig. **13**.

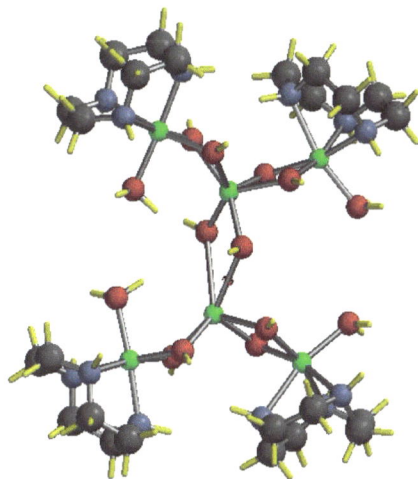

Figure 13: The two complexes are linked by oxygen bridges.

12. Referring to the structure shown in Fig. **13**, to the right and left of center, there are 3 oxygen atoms which resemble a triangle. The triangle on each side is made of one of the central O atoms, which bond Fe #2 and Fe #7, and the two protruding oxygen atoms which have two unoccupied bonding sites. On the left, the oxygen atoms are bonded to Fe #1 and Fe #6, and on the right the oxygen atoms are bonded to Fe #3 and Fe #8.

13. Add a Fe atom with an octahedral geometry to the oxygen atom on the left side (Fig. **13**) that bonds Fe #2 and Fe #7. Than connect two open bonding sites on the Fe atom to one of the open bonding sites on the O atoms attached to Fe #1 and Fe #6.

14. Repeat step 13 on the right side of the molecule. Use the other central oxygen atom binding Fe #2 and Fe #7, and the oxygen atoms with two bonding sites that are linked to Fe #3 and Fe #8. Once you minimize and save the structure, it should appear like that shown in Fig. **14**.

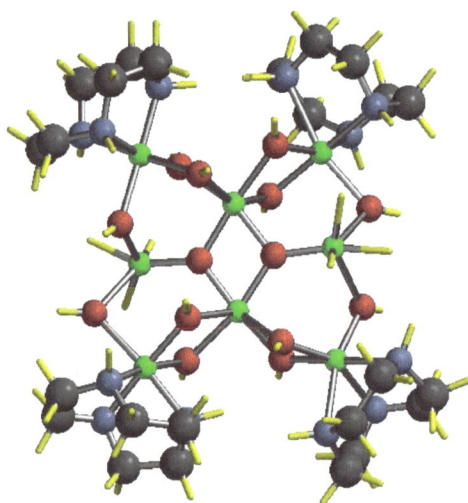

Figure 14: Two octahedral iron atoms are added.

15. The new Fe atoms just inserted are numbered Fe #4 (left), and Fe #5 (right). The nitrogen and carbon rings around Fe #4 and Fe #5 are the same ring as those around Fe #1, 3, 6, and 8. Begin to construct a ring

around Fe #4 and Fe #5 by adding an N atom with 4 bonding sites to each of the remaining bonding sites on Fe #4 and Fe #5.

16. Each of the N atoms added have three remaining bonding sites. On each N atom, add a sp^3, tetrahedral C atom to two of the three remaining bonding sites. Pick bonding sites closest to other nitrogen atoms. On each nitrogen atom there should be an open bonding site, pointing away from the other nitrogen atoms.

17. Construct a bond between a carbon atom on one nitrogen atom a carbon atom on another nitrogen atom. Repeat this step to form a ring on each side of the molecule. If made correctly it should be N-C-C-N-C-C-N-C-C looping back over itself. Spartan will automatically add hydrogen atoms to the carbons when the structure is minimized.

18. Your molecule is now complete, press the "V" button to view the overall shape, with H atoms added (see Fig. **15**). Run molecular mechanics on the structure and, under "Display" and "data" check that your empirical formula is correct. Be sure to save this structure in at least two locations, including an external memory device.

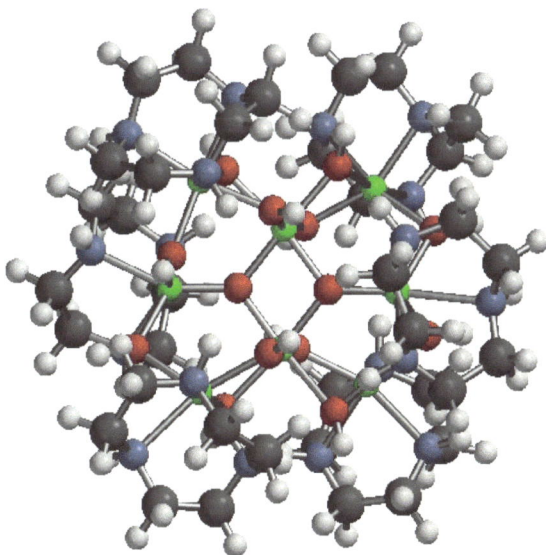

Figure 15: The completed single molecule magnet Fe₈.

19. Under the "Model" tab, you can change the appearance of your structure (example, see Fig. **16**).

20. Construct a table that will be used to record data measured on your Fe_8 structure (see Table **1**). This is only for data associated with iron atoms. Be sure to copy and paste your Fe_8 structure to your report. Copy/paste it from at least three different perspectives (3 images per page). Number the atoms in your image and correlate the numbers with the iron numbers in your Table **1**. Use arrows if needed. Be sure that each image has a figure caption.

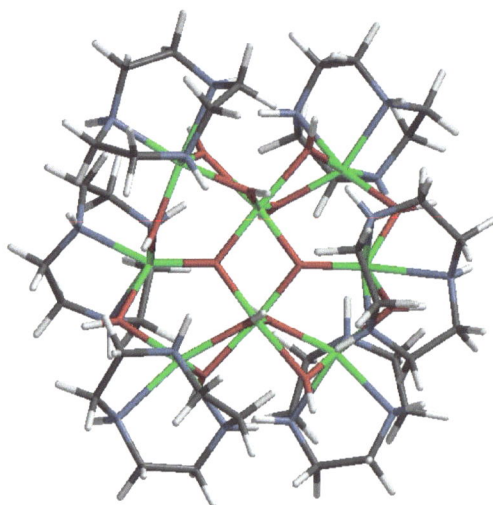

Figure 16: Fe_8 with a different appearance.

Table 1: An outline of a table that students fill in with parameters from their Fe_8 complex. The actual table would contain rows for all 8Fe atoms and their parameters.

Fe Atom #	Coordination Number	Ligand	Bond distance	Bond Angle from First Ligand
1	6	a. X b. Y . .	2.66 Å 2.51 Å 2.45 Å	0 65.4 77.3
2	6	a. X .	2.45	0

Exercise II. Building Mn$_{12}$

In this exercise the student will devise a method similar to that outlined with Fe$_8$ to construct the Mn$_{12}$ single molecule magnet (see Fig. **1**). First, number the Mn atoms from 1-12 on your 2-D image, in the order you will build the structure. After labeling the Mn atoms in a sequential fashion from one through twelve, begin by constructing structure in Spartan by only using Mn atoms and the oxide ions that connect them. Your first completed structural frame should be a Mn$_{12}$-oxide complex. Once this is complete, add the acetates and then the water molecules. It is easier to add molecular components (acetates, waters) than to build the structure atom by atom.

Like the Fe$_8$ exercise, copy/paste your structure (Spartan image) from at least three angles into your report, each with its own figure caption. Also, number the atoms and correlate this numbers (Mn #1, Mn #2, *etc.*) to the Mn numbers in the table you will build (see outline in Table **2**). Once complete, run your structure in molecular mechanics and check/confirm the empirical formula.

Table 2: An outline of a table that students fill in with parameters from their Mn$_{12}$-Ac complex. The actual table would contain rows for all 12 Mn atoms and their parameters.

Mn Atom #	Coordination Number	Ligand	Bond Distance	Bond Angle from First Ligand
1	6	c. H$_2$0 d. Acetate e. Oxide . .	(Mn-O) 2.56 Å 2.71 Å 2.89 Å	0 89.4 66.3
2	6			

Post Lab Questions

1. Attempt to run each structure using Molecular Mechanics in Spartan. What is the volume of a single Mn$_{12}$, Fe$_8$ cluster in Å^3? If your version of Spartan cannot handle this calculation estimate the length, width, and height and calculate V.

2. Does the acetate qualify as a ligand? A chelating agent? How about water or the oxide? Explain your answer.

REFERENCES

[1]   Linert, W.; Verdaguer, M. (Eds), *Molecular Magnets*, 1st ed., Springer-Verlag Wien, **2003**.
[2]   Gatteschi, D., Single molecule magnets: a new class of magnetic materials, *J Alloy Compd*, **2001**, *317–318*, 8–12.
[3]   Verdaguer, M. In: *Encyclopedia of Materials: Science and Technology*, 2nd ed., Elsevier, Oxford**, 2010**, pp. 1-9.

Send Orders of Reprints at reprints@benthamscience.net

CHAPTER 13

Ozone Decomposition Kinetics

Thomas J. Manning* and Aurora P. Gramatges

Department of Chemistry, Valdosta State University, Valdosta, Georgia, USA, and Instituto Superior de Tecnología y Ciencias Aplicadas, La Habana, Cuba

Abstract: In this exercise students will review some applications of an important industrial chemical. From this review it will become obvious why understanding basic physical and chemical parameters of chemical species are important. Students will also use existing experimental data, involving the decomposition of ozone to form oxygen, to determine reaction order and rate constant. Students will perform the calculations and graphing components of the exercise in a spreadsheet advancing their computational capabilities.

Keywords: Chemical kinetics, ozone, speciation, decomposition, industrial applications.

INTRODUCTION

Ozone (O_3) has found wide spread applications in society including aquaculture and aquarium water treatment, wastewater treatment, drinking water treatment, as a bleaching agent in the pulp industry, as a disinfectant, as an oxidizing agent in the chemical industry, treating swimming pool water, etching materials, and odor removal [1,2]. In most cases it is the combination of its strong reduction potential, its favorable environmental characteristics and the relative speed of its reactions with chemical and biological species that have seen its applications grow. The study of ozone formation in various types of discharges and plasmas has been an ongoing endeavor of scientists and engineers for both environmental and industrial applications.

Ozone can be produced by a variety of methods including corona discharge (CD), electrochemical cells, and UV light. The CD is the most common method for large-scale commercial production of ozone from pure oxygen or air. In the CD

---

*Address correspondence to **Thomas J. Manning**: Department of Chemistry, Valdosta State University Valdosta GA 31698, USA; Tel: 229-333-7178; E-mail: tmanning@valdosta.edu

several thousand volts are placed across two electrodes a few millimeters apart with the current flow regulated by the dielectric material. High temperatures and electron densities characterize this plasma or discharge medium.

Ozone has some well-known advantages over other strong oxidizing agents. Its product (O_2) is nontoxic when compared to the products of other oxidizing agents. For example, $HClO_2$ leaves behind a chlorine-based residue and fluorine gas (F_2) is highly corrosive. The kinetics of decomposition of a variety of organic compounds by ozone have been measured and shown to be quite favorable when compared to other strong oxidizing agents [3,4]. Ozone has equally impressive results as disinfection for such microorganisms as enterobacteria, viruses, bacterial spores, and amoebic Cysts in various water supplies. It regularly outperforms other common oxidizing agents such as HOCl, OCl⁻ and NH_2Cl in the inactivation of microorganisms and has the environmental advantage of leaving no residue.

Ozone's ability to absorb ultraviolet light in the 200 to 300 nanometer range and its subsequent depletion by chlorofluorocarbons has brought the oxygen allotrope much attention. The 1995 Nobel Prize in Chemistry was awarded to Mario Molina and F. Sherwood Rowland for their model predicting the effect that man-made chlorofluorcarbons (CFC's) have on ozone levels in the stratosphere. The first reaction of oxygen being transformed into ozone involves the dissociation of molecular oxygen (O_2) by ultraviolet light (hv) with a wavelength shorter than 240 nm to form atomic oxygen (O),

$$O_2 + hv \Rightarrow O + O$$

In the electrical discharge, electrons produce this dissociation of molecular oxygen.

$$O_2 + e^- \Rightarrow O + O$$

In viewing ozone production from a thermodynamic perspective, the conversion of oxygen to ozone is an endothermic process where energy (ΔH) is provided by the discharge,

$$3O_2(g) \Rightarrow 2O_3(g) \; \Delta H = +286 \text{ kJ}$$

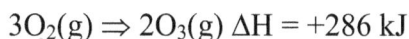

The atomic oxygen combines with molecular oxygen to produce ozone,

$$O(g) + O_2(g) \Rightarrow O_3(g)$$

The Gibbs Free energy (ΔG) change for the transformation of ozone back to oxygen is negative indicating that the reaction is thermodynamically favored,

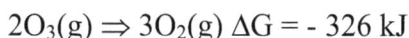

$$2O_3(g) \Rightarrow 3O_2(g) \; \Delta G = - 326 \text{ kJ}$$

In this portion of the lab the kinetics of this conversion of one allotrope of oxygen back to another form has already been measured and experimental data is provided to the student. The experimental data you are about to use involved the use of a Corona Discharge to dissociate O_2 and form O_3 (see equations above). Typically less that 1% of the O_2 entering the discharge is converted to O_3. Ozone is an unstable molecule and will typically decay back to oxygen in a matter of minutes or hours, depending on the conditions (*i.e.* pressure, temperature, catalyst present, *etc.*).

Once the O_3 is made the gas is transferred to a quartz cuvette. The cuvette, which is 10 cm long, is inserted in a spectrometer that measures the O_3 concentration. The ozone decomposition data you are provided with is ozone in pure O_2 or ozone in mixtures of O_2 and argon. Ar is 0.9% of the earth's atmosphere. Part of the exercise is to look at the impact that argon has on (1) the production of ozone in the discharge and (2) the impact that higher argon levels (and subsequently lower O_2 levels) have on the decomposition of ozone in the quartz cuvette under ambient conditions.

Pre-Lab Questions

1. Write out the time-concentration equations for zero, first and second order reactions. Define what each variable is in the equation and include the units.

2. Describe how a straight line plot is used with zero, first and second order data to obtain a rate constant (generate the plots for each in a 2D

drawing program, label the axis and illustrate the shape of the plot and how the slope is related to rate constant).

3.   List the equations to convert a rate constant to a half-life for a zero, first and second order reaction. Include units for all variables.

Table 1: (Exercise #1). Transfer the data below to a table in your spreadsheet. This data describe decomposition of ozone in pure a pure O_2 atmosphere (room temp and room pressure) and is measured over 75 minutes. Three separate experiments were conducted with ozone and oxygen trapped in a 10 cm quartz cell. The absorbance (Abs) values given here were measured at 254 nm. The four row here (time, abs, abs, abs) will be rows A, B, C, and D in your spreadsheet.

Time/mins	Abs	Abs	Abs
0	0.291	0.297	0.297
3	0.271	0.242	0.252
5	0.257	0.225	0.237
10	0.226	0.197	0.209
15	0.201	0.179	0.189
20	0.181	0.162	0.172
25	0.159	0.147	0.157
30	0.145	0.132	0.143
35	0.13	0.119	0.131
40	0.117	0.108	0.118
45	0.104	0.097	0.108
50	0.093	0.087	0.098
55	0.086	0.078	0.092
60	0.077	0.073	0.083
65	0.069	0.066	0.073
70	0.061	0.057	0.07
75	0.055	0.053	0.064

In this exercise the student will:

a.   Transfer all data from this exercise to a spreadsheet.

b.   Calculate the average absorbance (Abs) values of the three experiments in each chart. For example, looking at Table **1**, average 0.291, 0.297 and 0.297 and use this value in your Beers law calculation. This will be done for every row in every table.

c. Using Beers Law, the average absorbance values will be converted to average gas-phase concentrations of ozone (in Molar).

d. A first order plot (ln(conc)) *vs.* time) will be generated and the slope used to obtain the rate constant for the reaction. This will be done for all five data sets – you should have five first order plots, each with its own rate constant.

e. Each graph will have a figure caption in numerical order with a description of the system (*i.e.* % O_2, % Ar), its correlation coefficient and the rate constant for the systems.

f. Additional graphs will be made that examine the potential role that argon and/or oxygen play in the production of ozone in the discharge and the decomposition of ozone in the quartz cell.

The spreadsheet instructions below (in Excel) will take you through this analysis in a step by step fashion for the first data set. You should repeat this spreadsheet analysis for all five data sets.

1. Open a new worksheet in Excel.

2. Enter the header "time/mins" in location A1.

3. In location A2 enter the number "0".

4. In location A3 enter the number "3".

5. Enter the number "5" into location A4.

6. In location A5 enter the equation "=sum (A4+5)".

7. Copy and paste the equation down to location A18 (the value in A18 should be 75). This data set (A2.A18) are the time values (in minutes) from Table **1**.

8. Enter the header "Abs" in location B1. Copy the values (2nd row) from Table **1** into locations B2 through B18. This is your first set of

absorbance values monitoring the decomposition of ozone over 75 minutes.

9. Enter the header "Abs" into locations C1 and D1.

10. Enter the values from the 3rd and 4th row of Table **1** into locations C2-C18 and D2-D18. Save this table (which should have the same format as Table **1**).

Average and Standard Deviation

1. Enter the header "Avg Abs" into location E1.

2. Enter the equation "=Average (B2:D2)" in location E2.

3. Copy the equation into locations E3.E18.

4. Enter the header "Standard Deviation" into location F1.

5. Enter the equation "=STDEV(B2:D2)" in location F2 and copy/paste it down to F18.

6. Select the chart wizard icon in the toolbar of your spreadsheet. Select "XY(Scatter)" and click "Next".

7. Click the "Series" tab and select "Add".

8. Using the red arrowed-box select locations A2 through A18 for the "x values" and cells E2 through E18 for the "y values" and click "Next".

9. Enter "Change in Absorbance *vs.* Time/min" in the Chart Title Box. Also enter your name in the graph title.

10. Enter "Time(min)" in the values for x-axis and "Absorbance" in the values for y-axis.

11. Deselect "Show Legend" in the Legend Tab and Click "Next". With a single data set on a graph, there is no reason to indicate which series is listed.

12. Open the Chart as "a new sheet" and click "Finish". This graph should be copied to your report and, with the figure caption, should not take up more than ½ page.

Beer's Law is used to convert the absorbance (A), a unitless number, to concentration (c).

$$A = \varepsilon \, \iota \, c$$

Calculate the concentration of ozone in the gas phase where $\iota = 10$ cm path length, $\varepsilon = 3000$ $M^{-1}cm^{-1}$ (ε is referred to as the molar absorptivity or the extinction coefficient).

1. In location G1 enter the header "Average Concentration".

2. In cell G2 enter the equation "=sum(E2/(3000*10))". This is the equation rearranged to the form $A/(\varepsilon \, \iota)$.

3. Copy/paste the equation in location G2 from G3.G18.

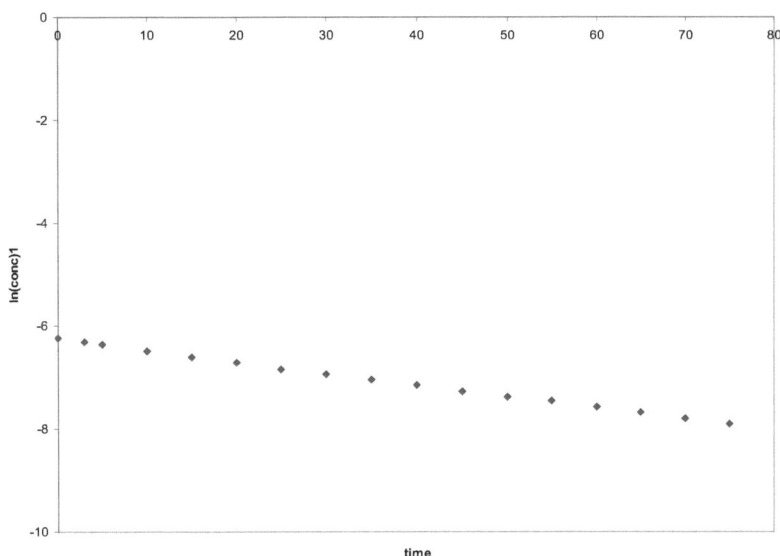

Figure 1: A first order plot for the decomposition of a chemical species. Note that the y-axis is decreasing negative numbers. Your plot should include a title, your name and units on the x-axis. It should also have the results of a linear best fit (y = mx+b equation and correlation coefficient) listed on the graph.

4. Enter the header "ln(conc)" in location H1.

5. In location H2 enter the equation "=ln(G2)". Copy/paste this equation from H3… H18.

6. Create a graph with the calculated natural log values (G2.G18), ln(conc.) on the y-axis, *vs.* time on the x-axis. Remember to include units, a title on the graph, and your name.

7. Using a linear fit, obtains the equation for the best fit straight line and displays the equation and the correlation coefficient within the boundaries of your graph (Fig. **1**). Copy the graph to your report and convert the slope of the line to a rate constant (1st order). Include the rate constant (with units!) in your figure caption.

8. Repeat steps 1-29 for the data sets in Tables **2**, **3**, **4**, and **5**. You should have five data sets each in their own spreadsheets.

9. At this point you should have plotted the first order data for the five data sets and obtained the rate constants (rate constants are always positive!). A sixth graph will now be generated using that data to see if argon impacted the decomposition of ozone in oxygen. Plot the rate constants (y-axis) *vs.* the oxygen concentration (1.0, 0.8, 0.6, 0.4, and 0.2). Copy the graph to your report and discuss the impact that argon had on the decomposition kinetics of ozone in an argon/oxygen environment.

10. Convert the first order rate constants to half-lives and plot the half-life (y-axis) *vs.* the oxygen concentration (x-axis) and explain the trend, if any exists.

Post-Lab Questions

Include the answers to these questions in your report after the graphs.

1. Plot the first ozone concentration (t=0) *vs.* the rate constant of each experiment. Does ozone play a role in its own decay? For example

does $O_3 + O_3 ==$ product or $O_3 + O_2 ==$ product appear to be the predominant mechanism?

2. Plot the starting ozone concentration verses the argon fraction. The starting or first ozone concentration indicates the concentration of the ozone being produced by the discharge that converts oxygen to ozone. Discuss, both qualitatively and quantitatively, what role argon plays a role in the production of ozone in an electrical discharge.

3. Explain the absorption of UV light by ozone, oxygen, argon, quartz and other plastics and how the selection of gases and materials is important in this experimental design.

Table 2: Experimental data measuring the decomposition of ozone in an environment of 20% Argon and 80% O_2 at room temperature and pressure. The ozone concentration is under 1% of the total gas pressure (O_2, Ar, O_3) so is not counted in the 20/80 assignment.

Time/mins	Abs (1)	Abs (2)	Abs (3)
0	0.681	0.677	0.685
3	0.568	0.573	0.581
5	0.537	0.558	0.562
10	0.481	0.518	0.524
15	0.445	0.487	0.494
20	0.413	0.457	0.466
25	0.383	0.429	0.44
30	0.364	0.402	0.415
35	0.334	0.371	0.39
40	0.305	0.343	0.369
45	0.285	0.321	0.347
50	0.264	0.297	0.326
55	0.251	0.29	0.307
60	0.239	0.271	0.289
65	0.222	0.254	0.272
70	0.206	0.234	0.254
75	0.19	0.219	0.235

Table 3: Experimental data measuring the decomposition of ozone in an environment consisting of 60% oxygen and 40% argon. The first value (Time = 0) is the value that indicates the concentration of ozone being generated by the corona discharge.

Time (min)	Abs	Abs	Abs
0	3.37	3.569	3.306
3	3.073	3.221	2.93
5	2.92	3.031	2.797
10	2.612	2.754	2.702
15	2.49	2.63	2.536
20	2.39	2.521	2.436
25	2.297	2.391	2.337
30	2.202	2.35	2.251
35	2.107	2.264	2.163
40	2.042	2.144	2.077
45	1.947	2.093	1.997
50	1.852	2.016	1.92
55	1.771	1.938	1.843
60	1.727	1.862	1.771
65	1.644	1.779	1.709
70	1.582	1.713	1.658
75	1.522	1.644	1.594

Table 4: Experimental data measuring the decomposition of ozone in an environment consisting of 40% oxygen and 60% argon. A quartz cuvette is used because quartz allows ultraviolet light to be transmitted whereas plastic or other glasses (*i.e.* Pyrex) will absorb UV light. The other gases involved in these experiments, argon and oxygen, absorb negligible amounts of UV. In this experiment only ozone readily absorbs UV radiation at 254 nm.

Time (min)	Abs	Abs	Abs
0	1.321	1.246	1.309
3	1.098	1.043	1.117
5	1.023	0.988	1.041
10	0.935	0.908	0.943
15	0.879	0.865	0.904
20	0.833	0.826	0.863
25	0.789	0.788	0.825
30	0.747	0.746	0.773

35	0.708	0.712	0.733
40	0.67	0.679	0.699
45	0.637	0.647	0.655
50	0.603	0.617	0.635
55	0.571	0.59	0.606
60	0.541	0.559	0.578
65	0.513	0.525	0.549
70	0.485	0.503	0.524
75	0.464	0.476	0.497

Table 5: Experimental data measuring the decomposition of ozone in an environment consisting of 20% oxygen and 80% argon. Time = 0 minutes represents the first measurement after the gas is collected from the corona discharge. The gas within the corona discharge can be several thousand degrees but rapidly cools to room temperature when transferred to a quartz cuvette.

Time (min)	Abs	Abs	Abs
0	2.082	2.175	1.89
3	1.793	1.894	1.67
5	1.674	1.78	1.556
10	1.501	1.559	1.388
15	1.411	1.482	1.314
20	1.339	1.426	1.256
25	1.275	1.357	1.205
30	1.203	1.297	1.151
35	1.169	1.233	1.1
40	1.112	1.187	1.05
45	1.059	1.141	0.991
50	1.013	1.078	0.973
55	0.963	1.034	0.936
60	0.902	0.973	0.888
65	0.864	0.947	0.856
70	0.841	0.907	0.826
75	0.8	0.865	0.789

REFERENCES

[1]   Beltran, F. J. Ozone Reaction Kinetics for Water and Wastewater Systems, CRC Press, Taylor and Francis, **2004**.

[2]   von Gunten, U. Ozonation of drinking water: Part I. Oxidation kinetics and product formation, Water Res., **2003**, 37, 1443-1467.

[3]   Lovato, M. E.; Martín, C. A.; Cassano, A. E. A reaction kinetic model for ozone decomposition in aqueous media valid for neutral and acidic pH, Chem. Eng. J., **2009**, 146, 486-497.

[4]   Buffle, M. O.; Schumacher, J.; Salhi, E.; Jekel, M.; von Gunten, U. Measurement of the initial phase of ozone decomposition in water and wastewater by means of a continuous quench-flow system: Application to disinfection and pharmaceutical oxidation, Water Res., **2006**, 40, 1884-1894.

CHAPTER 14

Thirty Equations for General Chemistry

Thomas J. Manning[*] and Aurora P. Gramatges

Department of Chemistry, Valdosta State University, Valdosta, Georgia, USA, and Instituto Superior de Tecnología y Ciencias Aplicadas, La Habana, Cuba

Abstract: In this exercise students will review the concepts and equations associated with thirty important relationships covered in general chemistry. Students will use a spreadsheet to graph correlations associated with each equation. This exercise may be given at any point in a general chemistry lab curriculum. If given at the beginning of a course than it can serve as an introduction to the course. If the exercise is assigned at the end of a course it serves as an excellent review for many of the concepts covered throughout a semester.

Keywords: Chemical concepts, equations, general chemistry, correlations.

INTRODUCTION

On the top of your report (typed entirely) will be your name, date and the project title. Each exercise has a graph that will be imported to your report along with questions to be answered. Each graph will have a title and your name (inserted in Spreadsheet program), axis's will be labeled (include units), and a numerical figure caption that briefly describes the data (*i.e.* Fig. **1**. This is a graph of....). If drawings are requested they will be done is a 2D drawing program. There should be a maximum of one graph (or one exercise) per page (your report will be a minimum of 30 pages long but some equations may require more than 1 page). Each exercise should start on the top of a new page. Pages should be numbered in the lower right hand corner and the final report will be stapled. Remember, no copying from web sources to answer questions, no exchanging of data, *etc.* Below are thirty relationships (equations) that are routinely used in various areas of chemistry and other areas of science [1-3]. In this exercise you'll apply these

---

Address correspondence to Thomas J. Manning: Department of Chemistry, Valdosta State University Valdosta GA 31698, USA; Tel: 229-333-7178; E-mail: tmanning@valdosta.edu

relationships to different systems that have some environmental or biomedical applications.

Each equation section below provides the name of the relationship (A), the equation (B), defines the variables and units (C), and a brief description of the concept (D). For your report, you'll create the graph and place a descriptive figure caption below it, using full sentences (part E). You'll also answer a question related to the chemical relationship. You may find the answer to this in your text, Wikipedia, *etc.* but your answer should be in your own words.

1. (A) Bohr's frequency condition (B) $E = h\nu$

(C) ΔE is the difference in energy (Joules)

h is the Planck's constant (6.626×10^{-34} J.s)

ν is the frequency (Hz) of the radiation

(D) Light energy has wave-particle duality. When the wavelength is short, the frequency is high and the photon has a high energy.

(E) 700 nm appears as red light, 600 nm as yellow light, 500 nm as green light and 400 nm as blue light. Convert each wavelength to a frequency and calculate its energy. Plot the wavelength (nm) *vs.* the energy (J, x-axis) and use a best fit line. What does the slope represent?

(F) Place the follow regions of electromagnetic radiation in order of (1) highest energy to lowest energy (2) shortest wavelength to longest wavelength (3) highest frequency to shortest frequency. Infrared (IR), ultraviolet (UV), gamma, microwaves, visible (VIS), x-ray, radio waves, vacuum ultraviolet (VUV).

2. (A) Frequency and Wavelength (B) $\lambda \nu = c$

(C) λ is wavelength of electromagnetic radiation (meters)

ν is frequency (Hz)

c is speed of light (3×10^8 m/s in a vacuum)

(D) Light or electromagnetic radiation is characterized by its frequency and wavelength, both of which are related to its energy. Frequency is the number of cycles that pass through a stationary point during a given period of time.

(E) 700 nm appears as red light, 600 nm as yellow light, 500 nm as green light and 400 nm as blue light. Convert each wavelength to a frequency and plot the wavelength verses the frequency (s^{-1}, x-axis) and apply a best fit line. What does the slope represent?

(F) the speed of light (c) is typically recorded as the speed of light in a vacuum. What is the speed of light in pure water? In a crystal such as diamond?

3. (A) Molarity equation (B) mol = MV

(C)   mol = moles

M = Molarity (moles/liters)

V = Volume (liters)

(D) The molarity is the concentration of solution as the number of moles of solute per liter of solution. A mole of any substance is defined as the amount of material containing 6.0221421×10^{23} particles.

(E) Considering you have one mole each of the following salts: sodium chloride, calcium chloride, magnesium chloride, iron (III) chloride, copper (II) chloride, tin (IV) chloride, phosphorous pentachloride, uranium (VI) chloride, tungsten (VI) chloride, manganese (VII) chloride. Plot the charge on the cation (x-axis) *vs.* the moles of chloride present.

(F) Find the molarity of the eight most common ions (cations/anions) found in blood with their respective concentrations.

4. (A) Molarity (B) M = mol/L

(C) molarity (moles/liter)

moles of solute (moles)

volume (liters)

(D) Molarity is a widely used concentration unit. It is abbreviated with a "M". For example, a label that says "6 M HCl" would read "six molar hydrochloric acid solution. This equation (#5) is an algebraic rearrangement of equation 4.

(E) A river flows into the ocean which results in freshwater changing into brackish water. A large part of the river is impacted by tides so saltwater from the ocean can be found several miles upriver during high tide. The units of concentration are parts per thousand (ppt) which is a mass percent measurement. 1 ppt is 1 gram of NaCl per 1000 milliliters (1000 grams) of water (D_{H2O} = 1 g/mL) or 1 mg of NaCl per 1 mL of water. A scientist measures this change of NaCl concentration along a river as it empties into the ocean:

Table 1: Salinity (salt concentration) data taken from a river that empties into the ocean.

Location (miles upriver)	ppt (NaCl)	Temperature
5 miles	1.1	25 °C
3 miles	3.05	25 °C
2 miles	11.5	25 °C
1 miles	18.9	25 °C
0.75 miles	25.9	25 °C
0.5 miles	31.8	25 °C
0.25 miles	33.2	25 °C
0.0 miles	35.0	25 °C

First convert the concentrations (Table **1**, ppt) to molarity. Assume that your 1.1 ppt solution is 1.1 grams of NaCl per 1 liter of water. 1.1 grams/(58.45 g/mol) results in the moles of NaCl in 1 liter of water. Plot the concentration of NaCl (in Molar) *vs.* the distance (x-axis) up river and use a linear (y = mx + b) fit to get the equation of the line. Put the equation and the correlation coefficient on your graph.

(F) List the ten most common ions (cations/anions) found in seawater and their concentrations in ppt and molarity.

5. (A) Dilution Equation (B). $M_1 V_1 = M_2 V_2$

(C) M_1V_1 is the initial molarity (moles/liters) and volume (liters) of the concentrated solution. M_2V_2 is the final molarity and volume of the diluted solution.

(D) This equation is often used to solve dilution problems in the aqueous phase.

(E) There is a starting solution of 100 milliliters of 35 ppt filtered seawater. You would like to use it as your stock solution for a conductivity curve but first you must make a series of dilutions (use RO water in the dilution). You want the final volume of each of your dilutions to be 10 mLs total. You want your final concentrations to be 5, 10, 15, 20, 25, 30 and 35 ppt. Plot the volume of your stock solution that you will use to make each 10 mL solution.

(F) Which solution has a higher concentration of salt solution, (1) a swimming pool of freshwater with a bucket of seawater added or (2) a table spoon of seawater? Which has a higher quantity of salt (in terms of grams)? Explain.

6. (A) Osmotic pressure (B) $\pi = iMRT$

(C) i = van't Hoff factor

M = Molarity (moles/liters)

R = Gas constant, where R = 0.08206 L atm $\cdot$ mol^{-1} $\cdot$ K^{-1}

T = Temperature (formerly called absolute temperature) (Kelvins)

(D) Osmosis is defined as the flow of solvent from a solution of lower solute concentration to one of higher solute concentration. Osmosis is a colligative property.

(E) You have five separate NaCl solutions (1.0 g/L, 3.0 g/L, 5.0 g/L, 10.0 g/L, 20.0 g/L). Calculate and plot the concentration of the salt (M, y-axis) *vs.* the pressure generated by these solutions across a semipermeable membrane at 25 $^\circ$C.

(F) Draw and label a simple osmotic cell (pure water on one side of membrane, 1.0 M sucrose on the other) and explain which way solutions flow across the membrane in a osmosis and reverse osmosis (RO) filter.

7. (A) Boyle's Law (B). $P_1V_1 = P_2V_2$

(C) P is pressure of a gas in a sealed system (P_1 is the initial pressure, P_2 is the final pressure).

V is volume (L) of the gas

(D) The units (*i.e.* atm, Pa, mmHg, Torr, psi, *etc.*) for P_1 and P_2 are not important BUT it is important they are both the same unit. The unit consistency is also critical for the pressure values. When the temperature is held constant, this is called isothermal.

(E) At the surface of the ocean there is 1 atmosphere of pressure exerted on your body. Once you are submerged, every 33 feet you descend the pressure on your body increases another 1 atmosphere. At 99 feet under the surface of the water there is 1 atm of pressure from the atmosphere and 3 atmospheres of pressure from the 99 foot water column or a total of 4 atm of total pressure is exerted on you. At 198 feet there would be approximately seven atmospheres of total pressure. Plot the total pressure (y-axis) exerted on you at 0, 33, 66, 99, 132, 165, 198 feet of depth.

(F) What are the bends (medical condition0 and how does Boyles Law help us understand this condition.

8. (A) Charles Law (B) $V_1/T_1 = V_2/T_2$

(C). V_1 = Initial Volume, V_2 = Final Volume

T_1 = Initial Temperature (K), T_2 = Final Temperature (K)

(D) Charles law shows the relationship between volume (V) and temperature (T) of a gas under isobaric (constant pressure) conditions in a sealed container. The volume units can vary (*i.e.* Liters, mL, *etc.*) must be consistent between V_1 and V_2. The temperature values must be in Kelvin (K).

(E) If you have an air bubble in the shape of a sphere (initial r = 0.1 mm) at 25 °C, plot its volume (x-axis) at 25, 30, 35, 40 and 45 °C (hint use radius to calculate volume of each sphere, also use K).

(F) Why can't Celsius be used for the temperature unit (hint what would happen at the freezing point of water in terms of math?)

9. (A) Combined Gas Law (B) $P_1V_1/T_1 = P_2V_2/T_2$

(C) P = Pressure

V = Volume

T = Temperature (K)

(D) This equation assumes that gas is trapped in a sealed system and two or three of the variables are altered. Boyle's law and Charles law as well as $P_1/T_1 = P_2/V_2$ can both be derived from this equality.

(E) A bubble of air starts at 198 feet where it is 0.5 mm in diameter. Plot the pressure (y-axis) exerted on it at the surface (28 °C), at 33 ft. below the surface (23.5 °C), at 66 ft. below (21.2 °C), at 99 ft. below (19.2 °C), at 132 ft. below (17.9 °C), at 165 ft. below (16.3 °C), and at 198 ft. below the surface (15 °C). Recall that the surface pressure is 1 atm and it increases 1 atm every 33 feet the depth increases. The temperature (K) should be on the z-axis of your three dimensional graph.

(F) if the number of moles in a sealed system changes during an expansion or a contraction does Charles Law, Boyles Law of the Combined Gas law still apply?

10. (A) Ideal Gas Law (B). PV = nRT

(C) P = pressure (atmospheres, atm)

V = volume (liters)

n = moles of gas (mol)

R = Gas law Constant (0.0821 liter. Atm/mol Kelvin)

T = temperature (Kelvin)

(D) The ideal gas law describes the relationship between the number of moles of a gas, its temperature and pressure, and the volume of the container holding it. It is a static system or one where there are no changes in P, V or T with time. Unlike the dynamic systems above (Boyle, Charles, *etc.*). The units listed above must be used.

(E) A 20 liter steel tank is filled to 2500 psi (14.7 psi = 1 atm) at 25 $^{\circ}$C with hydrogen gas. The hydrogen gas (H_2) is used to power a fuel cell. If the amount of hydrogen is reduced by 1 gram/hr for twenty hours from the tank under isothermal conditions, plot the moles of H_2 (x-axis) *vs.* the pressure every hour for twenty hours.

(F) What does the van der Waals equation correct for compared to the Ideal gas law?

11. (A) Law of Partial Pressure (B). $P_t = P_1 + P_2 + P_3....$

(C) P_t = total pressure of the gas mixture

P_1 = partial pressure of gas 1 (atm)

P_2 = partial pressure of gas 2 (atm)

P_3 = partial pressure of gas 3(atm)

(D) The law of partial pressure states the total pressure of a mixture of gases in the sum of the partial pressure of its individual components.

(E). Assume the percentage of gas in the atmosphere is the same as its partial pressure (N_2 = 78%, O_2 = 21%, Ar = 0.9%, CO_2 = 0.036%). Plot the molar mass of each species (x-axis) verses its partial pressure in the atmosphere. Is a correlation between molar mass and atmospheric composition.

(F). Will the partial pressure of the three major gas components be altered significantly by temperature? Explain.

12. (A) Henderson-Hasselbalch equation (B). $pH = pK_a + \log_{10}$ (base/acid)

(C) $pH = -\log [H^+]$ pH ranges from -1 to 14 an is a convenient measure of acidity

$pK_a = -\log K_a$ K_a is the equilibrium constant for a weak acid.

(D) The Henderson-Hasselbalch equation is used to estimate the pH of a buffer solution from the initial concentrations of the conjugate acid/base pair employed in the solution.

(E) For an acetic acid (K_a=1.8x10^{-5}) and acetate buffer solution, plot the pH (y-axis) as the acid concentration (0.05, 0.1, 0.2, 0.3, 0.4, 0.5) increases and the base (0.1 M) remains constant.

(F) Describe how bicarbonate (HCO_3^-) serves as a single system buffer in the ocean (pH=8.3) and human blood (pH = 7.34). Include relevant acidic and basic equations.

13. (A) Henry's law (B). $P_g = kC$

(C) P_g = <u>partial pressure</u> (atm) of the <u>solute</u> above the <u>solution</u>

C = <u>concentration</u> of the solute (gas) in the solution

k = Henry's Law constant, units might be L·atm/mol, atm/(mol fraction) or Pa·m^3/mol

(D) Henry's Law states the solubility of a gas in a liquid is proportional to its partial pressure in the gas phase above the surface, because an increase in pressure corresponds to an increase in the rate at which gas molecules strike the surface of the solvent.

(E) The Henry's Law constant for oxygen in water is 769.2 L·atm/mol. A rule of thumb for gases dissolved in water (lake, ocean, *etc.*) is that for every 33 feet below the surface the pressure increases by 1 atm. Calculate and plot the amount of oxygen in water (x-axis) *vs.* the pressure at sea-level/surface (1 atm), 33 feet (2 atm), 66 feet (3 atm), 99 (4 atm), 132 (5 atm), 165 (6 atm), and 197 (7 atm). Remember that O_2 is approximately 20% of air.

(F) What is the concentration of O_2 dissolved in human blood (look up). How does it compare to O_2 dissolved in water?

14. (A) Raoult's Law (B). $P = x_{solvent} P_{pure}$

(C) $x_{solvent}$ =   mole fraction of the component in solution

P_{pure} =   vapor pressure (torr) of the pure component

P   =   vapor pressure of the solvent in the mixture.

(D) Raoult's Law states the vapor pressure of a solvent in the presence of a nonvolatile solute if proportional to the mole fraction of the solvent in the mixture.

(E) The vapor pressure for water at 25 °C is 23.76 mmHg. If the mole fraction of water in a mixture decreases from 1.0 to 0.0 in 0.1 increments, plot the vapor pressure (y-axis) *vs.* the mole fraction.

(F) Which of the following solvents has the highest vapor pressure and the lowest vapor pressure (water, methanol, ethanol, carbon dioxide). Explain why in terms of intermolecular forces and the ability of the respective solvent to evaporate.

15. (A) Kinetics, First Order, Concentration Verses Time B. $\ln[A]_t = -kt + \ln[A]_0$

(C) $[A]_t$ = concentration of species A at time t

k   = rate constant (units of 1/t)

$[A]_0$ = initial or starting concentration

(D) A reaction in which only one molecule undergoes a chemical change is a first order reaction. The concentration or quantity units for A can be molar, moles, grams, ppm, *etc.* as long as both (A, A_0) are the same units.

(E) Assume you have a sealed container with 10^{-4} M ozone (O_3) at the start of the reaction. Its decomposition has a half-life of 90 minutes at a certain temperature and pressure. Plot the ozone concentration every five minutes over a five hour span.

(F) If ozone is an unstable molecule and decomposes to oxygen in a matter of minutes or hours, how can it continuously block ultraviolet light from the sun while in the stratosphere? (Hint, how is it made in the upper atmosphere)

16. (A) Kinetics, second order, concentration verses time (B). $1/[A]_t = kt + 1/[A]_0$

(C) $[A]_t$ = concentration of species A at time t

k = second order rate constant (units = 1/(time*conc))

$[A]_0$ = Initial or starting concentration

(D) A reaction in which two molecules react or collide to induce a chemical change. The units on the rate for a second order reaction are different than the units for a first order rate constant. The concentration or quantity units for A can be molar, moles, grams, ppm, *etc.* as long as both are the same and they match the units found in the rate constant (k).

(E) For a hypothetical reaction 2A➜ B the following data (time, conc) was obtained for the decomposition of A verses time (0 s,.0105 M), (61 s,.00679 M), (119 s,.0051 M), (182 s,.004101), (245 s,.00348M), (310 s,.00291), (360,.00262). Plot the 1/(conc A) (y-axis) *vs.* the time (axis) and use the graph to determine the second order rate constant.

(F) For the reaction A+B ➔ C, it has the second order rate constant k= 3.24 $s^{-1}M^{-1}$. If the reaction is second order with respect to A and zero order with respect to B, what is the rate law? Plot the rate of reaction (y-axis) *vs.* the starting concentration of A (axis), if six experiments had $[A]_o$ of 0.1, 0.075, 0.055, 0.040, 0.025, 0.0152 M and $[B]_o$ was held constant at 0.05 M. Explain what impact A has on the rate of reaction. On the same graph (2 series) plot the rate of reaction (y-axis) *vs.* the starting concentration of B (axis), if six experiments had $[B]_o$ of 0.1, 0.075, 0.055, 0.040, 0.025, 0.0152 M. Explain what impact A has on the rate of reaction $[A]_o$ was held constant at 0.05 M. Fit both data sets with its own best fit line and include the slope and correlation coefficient on the graph.

17. (A) Kinetics, first order, half-life (B). $t_{1/2} = 0.693/k$

(C) $t_{1/2}$ = half the time

k = rate constant

(D) For a first-order reaction, $t_{1/2}$ is independent of the initial concentration. The time units for $t_{1/2}$ and k have to be related. For example, if time is in seconds, than the rate constant is in 1/s or s^{-1} or if time is in years, than the rate constant will be in yrs^{-1}.

(E) The rate constants for six hypothetical reactions are 0.1 s^{-1}, 0.1 min^{-1}, 0.03 hr^{-1}, 1234.8 $days^{-1}$, 8.23 s^{-1} and 0.000482 ms^{-1}. Convert each to a half-life and change the units to in minutes. Than recalculate each of the rate constants in min^{-1}. Plot the rate constants (in min^{-1}) *vs.* the half-life (x-axis, in min) and include the graph in your report. Does the slope indicate any significant number?

(F) Naturally occurring nuclear reactions follow first order kinetics. Using the concepts outlined in defining unimolecular, bimolecular, and termolecular reactions, explain why the decay of a nucleus is a first order reaction.

18. (A) Kinetics, second order, half-life. (B). $t_{1/2} = 1/k[A]_o$

(C) $t_{1/2}$ = half life

k   = second order rate constant

$[A]_0$ = initial concentration

(D) For a second order reaction, the half-life depends on the initial concentration. Because a collision of two species is needed, the second order half-life increases as the concentration decreases. The second order rate constant has units of $1/(time*concentration)$, such as $M^{-1}s^{-1}$.

(E) A second order reaction has a rate constant of 9.1 $M^{-1}h^{-1}$ when the starting concentration $[A]_o$ is 0.1 M. Calculate the half-life for the reaction when the starting concentrations are 1.0 M, 0.75 M, 0.5 M, 0.25 M, 0.1 M. 0.075 M, 0.05 M, 0.025 M and 0.01 M. Plot the rate constant verses the half-life (axis) and include the equation of the best fit line and the correlation coefficient on the graph.

(F) For the reaction: 2 NO + O_2 -> 2 NO_2, the following results were obtained:

Experiment # [NO] [O_2] *Rate*

1   0.1 0.05 0.1

2   0.1 0.10 0.2

3   0.2 0.05 0.4

What is the order of the reaction with respect to [NO]? the overall rate law? The rate constant? Can the second order half-life equation be used to calculate the half-life? Explain?

19. (A) Arrhenius equation (B). $k = A^{-Ea/RT}$

(C) k = rate constant

R = the gas constant (8.314 J/mol.K)

A = constant called the frequency factor

E_a = activation energy (J)

T = temperature (K)

(D) The Arrhenius equation is sued to adjust the rate constant for a chemical reaction as a function of temperature. The activation energy is needed to conduct this calculation.

(E) A student runs a reaction at different temperatures (20 °C, 25 °C, 30 °C, 35 °C, 40 °C, 45 °C) and experimentally measures the first order rate constant at each temperature. The student uses the rate constant data to calculate the half-life at teach temperature (2.3 s, 3.42 s, 4.6 s, 5.79 s, 6.88s, 7.99 s). Expand the Arrhenius equation above using natural log (*i.e.* ln) and plot 1/T (in K^{-1}) *vs.* the ln$k$ and derive and report the rate constant.

(F) There is a form of the Arrhenius equation which does not have the "A" factor but docs have two rate constants (k_1 and k_2) and two temperatures (T_1 and T_2). Write out this equation and explain its use in modelling chemical kinetics. Also, what signs do activation energies, rate constants and temperatures used in the Arrhenius equation always possess?

(G) In a 2D drawing program, construct an energy diagram for the reaction of A➔B. It has an E_a of 40 kJ/mol without a catalyst and an E_a of 25 kJ/mol with a catalyst (include both in your diagram). Label its transition state (activated complex), its enthalpy (-12.3 kJ/mol), label each axis, and identify where the time/energies for the products and reactants.

20. (A) Nernst equation (B). $E_{cell} = E°_{cell} - 0.0592/ n \log Q$

(C) E_{cell} = cell potential (V)

$E°_{cell}$ = standard cell potential (V)

n = moles of electrons transferred (n = 1,2,3, *etc.*)

Q = reaction quotient (similar to K, equilibrium constant)

(D) An electrochemical potential ($E°_{cell}$) reaction assumes a temperature of 25 °C, a pressure of 1 atm and concentrations of 1 M for all species dissolved in a solvent. The Nernst equation allows for a correction to E^o_{cell} when the concentrations are not 1M.

(E) A common galvanic cell is the Danielle cell and can be represented with the notation

$$Zn(s) \mid ZnSO_4(aq) \parallel CuSO_4(aq) \mid Cu(s)$$

The two reduction half reactions for the cell are $Cu^{2+} + 2e^- \rightarrow Cu$ ($E^o = +0.34$ V) and $Zn^{2+} + 2e^- \rightarrow Zn$ ($E^o = -0.76$ V) and the total spontaneous reaction is $Cu^{2+} + Zn \rightarrow Cu + Zn^{2+}$ which results in an electric potential of $E^o_{cell} = +0.34$ V $-(-0.76$ V$) = 1.10$ V (note need one oxidation and one reduction reaction, hence the reversal of the Zn reaction and the sign switch). Using the Nernst equation, calculate the cell potential if the Cu^{+2} concentration decreases from 1.0 M to 0.1 M in 0.1 M increments (*i.e.* 1.0, 0.9, 0.8, …0.1). Plot the corrected cell potential (E) *vs.* the Cu^{+2} concentration (assume that Zn^{+2} are held at 1.0 M in all experiments).

(F) Explain the similarities and differences between Q and K in terms of the equilibrium constant (P/R) and reaction time.

(G) Using a 2D drawing program, outline and label a hydrogen fuel cell. Include reactions that take place at the anode and cathode.

21. (A) Measuring enthalpy (B). $\Delta H = mc\Delta T$

(C) $\Delta H =$ Change in enthalpy of a chemical system (J)

$m =$ mass of a system (g)

$c =$ specific heat (J/g. °C)

$\Delta T =$ temperature change (K)

(D) This equation is used in conjunction with experimental data. In a typical set up, a strong acid and a strong base (two typical reactants) are mixed in an insulated container. A temperature sensing device (thermistor, thermometer, *etc.*)

is used to measure the temperature change over time. This is ΔT and, knowing the specific heat of the solvent (*i.e.* H_2O = 4.184 J/g.°C) and the total mass of the solvent and reactants, the enthalpy can be easily determined.

(E) Listed in Table **2** is calorimetric data associated with a chemical reaction. Plot the data (time, x-axis) and measure the ΔT. The solvent is 100 mLs water (*i.e.* c = 4.184 J/g.°C; D_{H2O} = 1 g/mL). If 0.01 moles of A are reacted with 0.01 moles of B, what is the ΔH (in kJ/mol) for this reaction?

Table 2: Calorimetric data for an exothermic reaction (heat released). Note – data continued to next page.

Time (s)	Temp (°F)
0	71.6
10	71.6
20	71.6
30	71.6
40	71.6
50	71.6
60	71.6
70	71.6
80	71.6
90	71.6
100	71.6
110	71.6
120	71.6
130	71.6
140	72
150	73.5
160	75.2
170	76.4
180	77.5
190	78.3
200	78.7
210	79.1
220	79.5
230	79.8

240	79.9
250	80.1
260	80.2
270	80.3
280	80.4
290	80.45
300	80.5
310	80.5
320	80.55
330	80.6
340	80.6
350	80.6
360	80.6
370	80.65
380	80.65
390	80.65
400	80.65
410	80.65
420	80.65
430	80.65
440	80.65
450	80.65
460	80.65
470	80.65
480	80.65

(F) Compare the similarities and the differences between a solution calorimeter and a bomb calorimeter. List 2 similarities and 2 differences in their construction and operation.

22. (A) Gibbs Free energy, Enthalpy and Entropy. (B) $\Delta G = \Delta H - T\Delta S$

(C) ΔG = Gibbs Free energy change (J)

ΔH = Enthalpy change (J)

T = Temperature (K)

ΔS = Entropy change (J/K)

(D) A negative Gibbs Free energy is a spontaneous reaction (*i.e.* a battery), a positive is not spontaneous (N_2 + O_2 → 2NO under normal conditions); a negative enthalpy indicates a exothermic reaction (*i.e.* a combustion rxn) and a positive indicates an endothermic reaction (twist pack that turns cold); a positive entropy indicates a reaction where the disorder increases (ice melting to form water) and a negative entropy indicates a reaction where disorder decreases (water freezes to form ice).

(E) Over small temperature ranges and assuming no phase changes take place, one can assume that the shift in ΔH and ΔS due to temperature is negligible but ΔG can shift as the temperature varies. For an exothermic reaction (ΔH = -41.6 kJ/mol) in which the disorder increase (ΔS = 95.2 J/molK), calculate the free energy (in J) *vs.* the temperature (25 to 37 °C in 1 °C increments). Plot the results (Temp (K) on x-axis)

(F) For a phase transition (*i.e.* ice to water; water to steam), what is the value of ΔG and how does the equation outlined in this section appear when that value is inserted.

23. (A) Gibbs Free Energy and Redox potential (B). $\Delta G = -nFE_{cell}$

(C) ΔG = Gibbs Free Energy (J)

n = moles of electrons in balanced equation (*i.e.* 1,2,3.)

F = Faraday's constant (9.6485 x 10^4 C/mol)

E_{cell} = Cell potential (volts)

(D) Any oxidation and/or reduction reaction with a cell potential can also be described by thermodynamic parameters. Recall also that $\Delta G = \Delta H - T\Delta S$ so a simple substitution results in $-nFE_{cell} = \Delta H - T\Delta S$. Recall also that $\Delta G = -RTlnK$

(K = equilibrium constant) so another substitution yields $-nFE_{cell} = -RTlnK$, which show that equilibrium constants are related to redox potentials.

(E) In Table **3**, there is a list of weak acids, there equilibrium reaction and the equilibrium constant (K_a). In your spreadsheet, plot the K_a (x-axis) for each acid verses its Gibbs free energy (in kJ/mol). Plot the Gibbs energy on the first (left) y-axis. Than plot the K_a *vs.* the cell potential (cell potential on second axis, right). Use a best fit for each data set. What do the slopes indicate in each plot.

(F) If Redox potentials fall over the +3.0 V to -3.0 V range, what are the normal limits for Gibbs Free energy and equilibrium constants using these boundaries.

Table 3: Some common weak acids, their equilibrium expression and their equilibrium constants in the aqueous phase.

Acetic acid (HAc, in vinegar)	$HC_2H_3O_2 \leftrightarrows H^+ + C_2H_3O_2^-$	1.8×10^{-5}
Benzoic acid (food preservative)	$C_6H_5CO_2H$	
	$\leftrightarrows H^+ + C_6H_5CO_2^-$	6.4×10^{-5}
Chlorous acid (in pure form it is unstable)	$HClO_2 \leftrightarrows H^+ + ClO_2^-$	1.2×10^{-2}
Formic acid (ant bites!)	$HCHO_2 \leftrightarrows H^+ + CHO_2^-$	1.8×10^{-4}
Hydrocyanic acid (CN^- is cyanide)	$HCN \leftrightarrows H^+ + CN^-$	6.2×10^{-10}
Hydrofluoric acid (note HCl, HBr, HI are strong acids)	$HF \leftrightarrows H^+ + F^-$	7.2×10^{-4}
Hypobromous acid	$HOBr \leftrightarrows H^+ + OBr^-$	2×10^{-9}
Hypochlorous acid (bleach!)	$HOCl \leftrightarrows H^+ + OCl^-$	3.5×10^{-8}
Hypoiodous acid	$HOI \leftrightarrows H^+ + OI^-$	2×10^{-11}
Lactic acid (builds up during exercise)	$CH_3CH(OH)CO_2H$	
	$\leftrightarrows H^+ + CH_3CH(OH)CO_2^-$	1.38×10^{-4}
Nitrous acid (important atmospheric intermediate)	$HNO_2 \leftrightarrows H^+ + NO_2^-$	4.0×10^{-4}
Phenolic acid	$HOC_6H_5 \leftrightarrows H^+ + OC_6H_5^-$	1.6×10^{-10}
propionic acid (a carboxylic acid)	$CH_3CH_2CO_2H$	
	$\leftrightarrows H^+ + CH_3CH_2CO_2^-$	1.3×10^{-5}

(G) The following are seven common strong acids (HCl, hydrochloric acid; HBr, hydrobromic acid; HI, hydroiodic acid; H_2SO_4, sulfuric acid (first proton only is strong; HNO_3, nitric acid; $HClO_4$, perchloric acid; $HClO_3$, chloric acid) and two common strong bases (NaOH, sodium hydroxide; potassium hydroxide). When these species dissociate to form cations and anions in solution, is there chemical equilibrium (equilibrium constant?) or does the reaction generate a potential? Explain.

24. (A) Boiling Point Elevation (B) $\Delta T_B = K_b m$

(C) ΔT_B = Temperature increase of boiling point ($^\circ$C)

K_b = Boiling Point elevation constant ($^\circ$C/m)

m = molality (moles solute per kilogram solvent)

(D) Adding a substance such as a salt (*i.e.* NaCl) raises the boiling point of a solvent (*i.e.* H_2O). The concentration unit used is molality, which is calculated by taking the moles of solute and dividing by the kilograms of solvent (*i.e.* moles/kg).

(E) NaCl is added to 100 mLs of water in ten 0.5 gram increments (00.5, 1.0, 1.5, …5.0 g). Calculate the new boiling point (hint, start at 100 $^\circ$C) at each addition and plot the boiling point (y-axis) *vs.* the total mass of NaCl added (0, 0.5, 1.0, 1.5,…).

(F) Briefly explain how molarity, normality and molality are different units of concentration.

25. (A) Freezing Point Depression (B) $\Delta T_F = K_f m$

(C) ΔT_F = decrease in freezing point temperature ($^\circ$C).

K_f = freezing point depression of solvent. ($^\circ$C/m)

m = molality of salt in solvent (mol/ kg)

(D) Adding a salt to a solution will lower its freezing point. For example, salt is spread on roads to melt ice, it drops the freezing point of water below 0 °C.

(E) NaCl is added to 100 mLs of water in ten 0.5 gram increments (0, 0.5, 1.0, 1.5, ...5.0 g). Calculate the new freezing point (hint, start at 0 °C) at each addition and plot the boiling point (y-axis) *vs.* the total mass of NaCl added (0, 0.5, 1.0, 1.5,...).

(F) Define the term "colligative property" and list the four examples.

26. (A) Rydberg Equation (for hydrogen atom) (B). $1/\lambda_{vac} = R_H \, Z^2 \, (1/\eta_1^2 - 1/\eta_2^2)$

(C) λ_{vac} = wavelength (meters) of photo emitted in vacuum,

R_H = Rydberg Constant for Hydrogen ($1.097 \times 10^7 \, m^{-1}$)

η_1 = lower energy level

η_2 = higher energy level ($\eta_1 < \eta_2$; both are positive integers 1,2,3,4, *etc.*)

Z = atomic number (1 for hydrogen).

(D) Because the hydrogen atom has a single electron, it can be fairly easily modeled with the Rydberg formula. When atoms have more than one electron, the complexity of modelling the energy levels increases dramatically.

(E) For the following six transitions in a hydrogen atom ($\eta_1 \Rightarrow \eta_2, \eta_1 \Rightarrow \eta_3, \eta_1 \Rightarrow \eta_4, \eta_1 \Rightarrow \eta_5, \eta_1 \Rightarrow \eta_6, \eta_1 \Rightarrow \eta_7$), calculate the λ_{vac} and plot the wavelength (y-axis) *vs.* the difference ($\eta_4 - \eta_1 = 3$) in energy levels.

(F) What are the Lyman, Balmer, Paschen, Brackett, Pfund, and Humphreys series and how do they relate to the Rydberg formula.

27. (A) de Broglie Equation (B) $\lambda = h/mv$

(C) λ = wavelength (meters)

h = Planck's constant (6.626 x 10^{-34} J.s)

m = mass (kg)

v = velocity (m/s)

(D) The wavelength of an electron (λ) of mass (m) moving at velocity (v) is represented by the de Broglie relation. Any object with a mass and velocity has a wavelength!

(E) An electron has a mass of 9.11×10^{-31} kg. Plot is wavelength (y-axis) verses its velocity if its velocity is 0.001 %, 0.005%, 0.0076 %, 0.01 %, 0.026%, 0.052%, 0.091%, 0.1%, 0.23%, 0.65 % and 1 % the speed of light.

(F) Briefly (6-7 sentences) describe the Davisson-Germer experiment (1927, Bell Labs, NJ) and how it complemented the Bragg experiment with x-rays and helped prove the de Broglie hypothesis.

28. (A) Clausius-Clapeyron equation (B) $\ln(P_2/P_1) = -\Delta H_{vap}/R.(1/T_2 - 1/T_1)$

(C) P = vapor pressure (torr)

ΔH_{vap} = Heat of vaporization (J/mol)

T = temperature (K)

R = Gas constant (8.314 J/mol.K)

P_1 is the vapor pressure at T_1, and P_2 at T_2.

(D) The relationship between vapor pressure and temperature is exponential and can be predicted or modeled if the heat of vaporization is known.

(E) The vapor pressure for water at 15 °C is 12.79 mmHg and 23.76 mmHg at 25 °C. First calculate the heat of vaporization for water. Once you have this value, generate a graph (use spreadsheet) that plots the vapor pressure for water (y-axis) *vs.* the temperature. Use temperature values from 1 °C to 99 °C in 1 degree increments.

(F) For water, compare the magnitude of the heat of vaporization to the heat of fusion. Which is greater and explain why one is bigger than the other?

29. (A) Density (B) D=M/V

(C) D = density (kg/L or g/mL)

M = mass (kg or g)

V = volume (L or mL)

(D) Density can be used to describe a solid, liquid, gas, a supercritical fluid or a plasma.

(E) Using the data in the table below, plot the density (x-axis) of the metal verses its Z# (# protons) and the density (a-axis) *vs.* the molar mass of the metals. Plot of best fit line and include the equation and correlation coefficient on the graph. Is there a correlation between density and either parameter? Explain?

Table 4: The densities of several metals.

Metal	Density (g/cm^3)
aluminum	2.70
zinc	7.13
iron	7.87
copper	8.96
silver	10.49
lead	11.36
mercury	13.55
gold	19.32

(F) In most cases density will be used to describe a solid, liquid or gas sample but can be used for two other phases, plasma and supercritical fluids. Describe physically/chemically each of these and give an example of each phase.

30. (A) Electrostatic equation (B). $F = q_1q_2/(4\pi\varepsilon_0r^2)$

(C) F = Force of attraction or repulsion (Newton's)

ε_o = Permittivity of free space

q_1, q_2 = charge magnitudes

r = distance between charged species (m)

(D) For two particles with the same charge (*i.e.* 2 electrons, -1 and -1) this equation calculates a force of repulsion. For two particles with opposite charges (Na^+, Cl^-), this calculates an attractive force. It assumes the two species are in a vacuum.

(E) For two ions (Na^+, Cl^-) are 100 nm apart in a vacuum. Ignore all terms in the equation except F and r^2. ($F = k/r^2$, assume k =1). As the two ions come closer (100, 99.9 nm, 99.8 nm, 99.7 nm,....0.2, 0.1 nm) calculate the F at each 0.1 nm increment down to zero nm. Plot the relative force of attraction (y-axis) *vs.* the distance (x-axis).

(F) Considering the calculation just conducted and plotted (Part E), how might this change if the two ions were in water?

REFERENCES

[1]   Ebbing, D.; Gammon, S. D. *General Chemistry*, 10th ed., Brooks Cole, **2012**.
[2]   Harré, R. *Laws in Chemistry, Philosophy of Chemistry*, North-Holland, Amsterdam, **2012**.
[3]   Warren, W. S. *Essential Physical Concepts for Chemistry, The Physical Basis of Chemistry,* 2nd ed., Academic Press, San Diego, **2000**.

Send Orders of Reprints at reprints@benthamscience.net

CHAPTER 15

First Order Kinetics and Naturally Occurring Radioactivity

Thomas J. Manning* **and Aurora P. Gramatges**

Department of Chemistry, Valdosta State University, Valdosta, Georgia, USA, and Instituto Superior de Tecnología y Ciencias Aplicadas, La Habana, Cuba

Abstract: In this exercise students will learn about radioactive decay schemes and isotopes. Students will simulate the quantity of several isotopes present as a function of time.

Keywords: Chemical kinetics, natural radioactivity, radioactive decay schemes, half-life.

INTRODUCTION

Radioactivity is a natural process that can be measured in soil and water. Natural decay schemes are found in nature and account for the presence of many of the heavier elements and their isotopes [1-3]. There are four well known natural decay schemes that have been identified by scientists including the Neptunium series (Table **1**), the thorium series (Table **2**), the radium series (Table **3**) and the actinium series (Table **4**). As you examine these tables you will notice that the half-lives vary tremendously. Some isotopes exist for a fraction of a second while others stick around for millions of years. Naturally occurring nuclear decay kinetics typically follows first order kinetics. In this exercise the student will simulate the quantity of several isotopes present as a function of time.

In conducting this exercise on Excel, there are some limitations to the spreadsheet that must be outlined. First, a typical spreadsheet can handle approximately 32,000 points. While this may appear to be a large number, let's assume you had three species you wanted to plot. A ($t_{1/2}$ = 5 million year), B ($t_{1/2}$ = 1 millisecond) and C($t_{1/2}$ = 5 million years) and they disintegrate A => B => C. In order to plot

---

***Address correspondence to Thomas J. Manning:** Department of Chemistry, Valdosta State University Valdosta GA 31698, USA; Tel: 229-333-7178; E-mail: tmanning@valdosta.edu

the millisecond species as a function of time you might need a point or concentration every 0.2 millisecond. In order to plot the other two species you'd need a point every 1 million years. Estimate how many times 0.2 milliseconds goes into 1 million years? Clearly you cannot do both of these species in a spreadsheet and have enough resolution to identify the species A, B and C. Given this variation in half-lives it would be difficult to plot all of the species in a radioactive decay scheme.

Table 1: An outline of the Neptunium decay series.

Nuclide- Product of Decay	Decay Mode	Half Life
^{232}Th-^{228}Ra	Alpha	$1.405 \cdot 10^{10}$ years
^{228}Ra-^{228}Ac	Beta	5.75 years
^{228}Ac-^{228}Th	Beta	6.25 hours
^{228}Th--^{224}Ra	Alpha	1.9116 years
^{224}Ra-^{220}Rn	Alpha	3.6319 days
^{220}Rn-^{216}Po	Alpha	55.6 second
^{216}Po-^{212}Pb	Alpha	0.145 second
^{212}Pb-^{212}Bi	Beta	10.64 hours
^{212}Bi-^{212}Po,^{208}Tl	Beta 64.06% Beta 35.94%	60.55 min
^{212}Po-^{208}Pb	Alpha	299 ns
^{208}Tl-^{208}Pb	Beta	3.053 min
^{208}Pb	.	stable

Table 2: The Thorium decay series.

Nuclide-Reactants and Decay Products	Decay Mode	Half Life
^{241}Pu-^{241}Am	Beta	14.4 years
^{241}Am-^{237}Np	Alpha	432.7 years
^{237}Np-^{233}Pa	Alpha	$2.14 \cdot 10^{6}$ years
^{233}Pa-^{233}U	Beta	27.0 days
^{233}U-^{229}Th	Alpha	$1.592 \cdot 10^{5}$ years

Table 2: contd….

^{229}Th-^{225}Ra	Alpha	$7.54 \cdot 10^4$ years
^{225}Ra-^{225}Ac	Beta	14.9 days
^{225}Ac-^{221}Fr	Alpha	10.0 days
^{221}Fr-^{217}At	Alpha	4.8 m
^{217}At-^{213}Bi	Alpha	32 millisecond
^{213}Bi-^{209}Tl	Alpha	46.5 min
^{209}Tl-^{209}Pb	Beta	2.2 min
^{209}Pb-^{209}Bi	Beta	3.25 hrs
^{209}Bi-^{205}Tl	Alpha	$1.9 \cdot 10^{19}$ years
^{205}Tl	.	Stable

Table 3: The Radium decay series.

Nuclide- Product of Decay	**Decay Mode**	**Half Life**
^{238}X-^{234}Th	Alpha	$4.468 \cdot 10^9$ years
^{234}Th-^{234}Pa	Beta	24.10 days
^{234}Pa-^{234}U	Beta	6.70 hours
^{234}U-^{230}Th	Alpha	245500 years
^{230}Th-^{226}Ra	Alpha	75380 years
^{226}Ra-^{222}Rn	Alpha	1602 years
^{222}Rn-^{218}Po	Alpha	3.8235 days
^{218}Po-^{214}Pb,^{218}At	Alpha 99.98 % Beta 0.02 %	3.10 min
^{218}At-^{214}Bi,^{218}Rn	Alpha 99.90 % Beta 0.10 %	1.5 sec
^{218}Rn-^{214}Po	Alpha	35 millisecond
^{214}Pb-^{214}Bi	Beta	26.8 min
^{214}Bi-^{214}Po,^{210}Tl	Beta 99.98 % Alpha 0.02 %	19.9 min
^{214}Po-^{210}Pb	Alpha	0.1643 millisecond
^{210}Tl-^{210}Pb	Beta	1.30 min
^{210}Pb-^{210}Bi	Beta	22.3 years
^{210}Bi-^{210}Po,^{206}Tl	Beta 99.99987% Alpha 0.00013%	5.013 days
^{210}Po-^{206}Pb	Alpha	138.376 days
^{206}Tl-^{206}Pb	Beta	4.199 min
^{206}Pb	.	Stable

Table 4: The Actinium decay series.

Nuclide	Decay Mode	Half Life
^{239}Pu-^{235}U	Alpha	$2.41 \cdot 10^4$ years
^{235}U-^{231}Th	Alpha	$7.04 \cdot 10^8$ years
^{231}Th-^{231}Pa	Beta	25.52 hours
^{231}Pa-^{227}Ac	Alpha	32760 years
^{227}Ac-$^{227}Th,^{223}Fr$	Beta 98.62% Alpha1.38%	21.772 years
^{227}Th-^{223}Ra	Alpha	18.68 days
^{223}Fr-^{223}Ra	Beta	22.00 min
^{223}Ra-^{219}Rn	Alpha	11.43 days
^{219}Rn-^{215}Po	Alpha	3.96 sec
^{215}Po-$^{211}Pb,^{215}At$	Alpha 99.99977% Beta 0.00023%	1.781 millisecond
^{215}At-^{211}Bi	Alpha	0.1 millisecond
^{211}Pb-^{211}Bi	Beta	36.1 min
^{211}Bi-$^{207}Tl,^{211}Po$	Alpha 99.724% Beta 0.276%	2.14 min
^{211}Po-^{207}Pb	Alpha	516 millisecond
^{207}Tl-^{207}Pb	Beta	4.77 min
^{207}Pb	.	Stable

In the first part of the exercise we will look at the decay sequence

$$^{213}Bi => {}^{209}Tl => {}^{209}Pb => {}^{209}Bi$$

which is part of the Neptunium series (Table **1**)? Note that the half-lives are on similar time scales (46.5 minutes, 2.2 minutes, 3.25 hours).

1. Open a new excel sheet and label column 1 (position A1) "Time, minutes".

2. In A2 place the value "0" and in A3 type the equation "=SUM(A2+0.1)".

3. Copy and paste this equation down to position A10000.

4. In B1 type the heading "starting mass" and in B2 enter the number "100" and copy it down to B10000. 100 grams is your starting mass for the Bi-213.

5. For the header of column C enter "k, Bi-213". In C2 calculate the rate constant k (in min^{-1}) using the equation " =SUM(0.693/45.6) ", which is from the first order half-life calculation: k = 0.693 / $t_{1/2}$. Copy the equation down to C10000. The value 0.015197 should appear in each box.

6. In D1 type the header "Mass A overtime". This column will use a rearrangement of the first order concentration verses time relationship:

    $$ln(A/A_o) = - kt$$

    In box D2 enter the equation "=EXP(-1*C2*A2+LN(100))" and copy it down to D10000. It should start with the calculated value of 100 and steadily decrease but always be a positive number.

7. In location E1 enter the title "Mass of Tl-209" and in E2 enter the equation "=SUM(100-D2)" and copy it down to E10000. It should start at zero grams but you may see an exceeding low number (*i.e.* -10^{-15}) due to rounding off numbers.

8. In location F1 enter the header "k, Tl-209" and in F2 calculate the rate constant for the decay of Tl using the equation "=SUM(0.693/2.2)". The value 0.315 should appear. Copy the equation down to F10000.

9. In G1 type the header "Mass, Tl, over time" and in G2 enter the equation "=EXP(-1*F2*A2+LN(E2)) ". This is the first order concentration verses time equation (eq. 2) rearranged. Copy this equation down to G10000. In the first slot (G2) the error message (#NUM) may appear. This is due to taking the natural log of the

negative round off error in E2. From G3 on the values should increase to G33 and then start a gradual decrease.

10. In location H1 type the header "Mass Pb-209, Start" and in location H2 enter the equation: " =SUM(100-G2-D2) ". Copy this equation down to position H10000.

11. In location I1 type the header "Rate Constant, Decay, Pb-209". In location I2 enter the equation "=SUM(0.693/205)", which is converting the half-life to the first order rate constant. Copy this equation down to location I10000.

12. In location J1 type the header "Mass. Pb-209 over time". In J2 enter the equation "=EXP(-1*I2*A2+LN(H2)) " This is a rearrangement of the equation 2, the first order concentration verses time equality. Copy this equation down to J10000.

13. In location K1 enter the header "sum of 3 masses; Bi-213, Tl-209, Pb-209" and in location K2 type the equation "=SUM(D2+G2+J2)" and copy it down to K10000.

14. In location L1 enter the header "mass Bi-209, stable isotope". ^{209}Bi has a half-life on the order of 10^{19} years so relative to the other three isotopes modeled in this graph, it is considered stable. In location L2 enter the equation, "=SUM(100-K2)" and copy it down to L10000.

You've completed all of the calculations needed for your graph. We will now graph them. For this graph your x-axis will be labeled "Time (min)" and your y-axis will be labeled "Mass (g)". Your x-axis should span from 0-1000 minutes and your y-axis should span from 0-100 grams. You will have four series:

1. A2…….A10000 (x-axis) verses D2……D10000 (y-axis). For the legend, name this series "Bi-213".

2. A2…….A10000 (x-axis) verses G2……G10000 (y-axis). For the legend, name the series "Tl-209".

3. A2........A10000 (x-axis) verses J2....J10000 (y-axis). For the legend, name the series "Pb-209"

4. A2........A10000 (x-axis) verses L2....L10000 (y-axis). For the legend, name the series "Bi-209".

Plot the graph and copy/past it to your report. It should have an appearance similar to the graph shown in Fig. **1**. Notice that you cannot see the Tl-209 species on your graph. This is NOT a mistake. Change the scale on the x-axis to 0-35 minutes and the scale on the y-axis to 0-4 grams. You should be able to see Tl-209 at this point. With a half-life shorter than the other species, it does have quite the presence of the other species with longer half-lives.

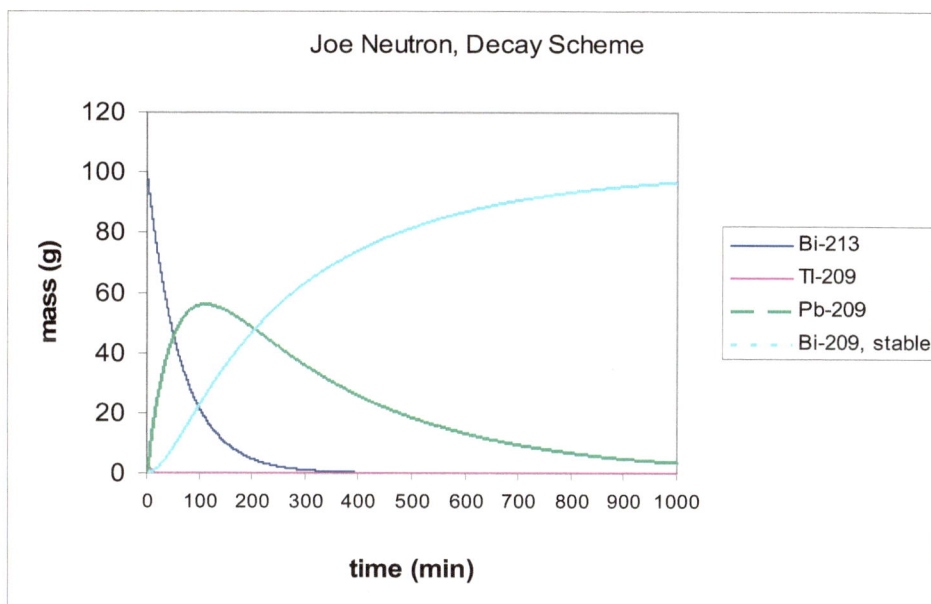

Figure 1: A nuclide speciation plot involving four species found in the Neptunium series.

Additional Assignment

Choosing from the other three decay schemes, pick three consecutive decay products from one of the radioactive series. The species should have half-lives within two orders of magnitude of each other. Once your series and a species are selected, generate a nuclide speciation plot like the one shown Fig. **1**. The third species might be very stable or have a short half-life, but only plot its increase in mass over time (ignore the impact of its decay and the formation of a fourth species). Also, assume you have 100 grams of starting material.

REFERENCES

[1] İpek, U.; Öbek, E.; Akca, L.; Arslan, E. I.; Hasar, H.; Doğru, M.; Baykara, O. Determination of degradation of radioactivity and its kinetics in aerobic composting, *Bioresource Technol.*, **2002**, *84*, 283-286.

[2] Ciffroy, P.; Garnier, J. M.; Pham, M. K. Kinetics of the adsorption and desorption of radionuclides of Co, Mn, Cs, Fe, Ag and Cd in freshwater systems: experimental and modelling approaches, *J. Environ. Radioactiv.*, **2001**, *55*, 71-91.

[3] Le, N. A.; Ramakrishnan, R.; Dell, R. B.; Ginsberg, H. N.; Brown, W. B. In: *Methods in Enzymology*, Academic Press, **1986**, Volume 129, pp. 384-395.

Send Orders of Reprints at reprints@benthamscience.net

Computer Based Projects for a Chemistry Curriculum, 2013, 163-182 163

CHAPTER 16

Chemistry in a Nanodrop: From H-Bonds to Peptides

Thomas J. Manning* and Aurora P. Gramatges

Department of Chemistry, Valdosta State University, Valdosta, Georgia, USA, and Instituto Superior de Tecnología y Ciencias Aplicadas, La Habana, Cuba

Abstract: In this exercise students will replicate interactions in a simulated solvent (water, ethanol, *etc.*) drop that have diameters of a few nanometers across. Students will construct systems that will allow them to review some fundamental interactions such as hydrogen bonding, dipole-dipole interactions and ion-dipole interactions. Students will search for correlations or trends involving molecular interactions and physical parameters. Students will look at ionic, atomic and small molecule interactions in a nanodrop of a common solvent such as water, ethanol and methanol. The diameter of these nanodrops typically ranges from two to six nanometers. Students will examine the interaction of a solvent nanodrop with two enkephalin peptides to better understand molecular folding.

Keywords: Solvents, chemical bonds, molecular interactions, correlations, nanomaterials.

INTRODUCTION

In experimental exercises students induce, measure and observe chemical reactions on a macroscale. When salt is added to water and stirred, it is a visual observation that determines whether it is soluble or insoluble. These observations are made from the collective actions of a very large number of ions, atoms and molecules behaving in a similar manner. In this exercise students will replicate some of these interactions on a nanolevel to better understand their macro observations [1-3].

Pre-Lab Exercises

1. Calculate the volume of sphere with a diameter of 4.05 nm. Calculate the diameter in nm^3 and cm^3.

2. What is the mass, in grams, of 1000 water molecules?

---

*Address correspondence to **Thomas J. Manning:** Department of Chemistry, Valdosta State University Valdosta GA 31698, USA; Tel: 229-333-7178; E-mail: tmanning@valdosta.edu

3.   If there are 1000 water molecules contained in a sphere of diameter of 4.05 nm, what is the density in grams/cm^3.

4.   The density of water is approximately 1.0 grams/cm^3. (answer each question in a bullet format)

  •   Would you expect a nanodrop of water to have the same density as a cup of water? Explain.

  •   Would other values such as vapor pressure and surface tension be the same for the macro and nano volumes of water or any other solvent? Explain.

  •   Is the ratio of surface area to total volume (SA/V) on a nanodrop the same as SA/V ratio of water in a beaker or a test tube? Explain.

Exercise (Instructions)

A.   For the first set of structures, use the following instructions for computational work, perform Single Point Energy; Semiempirical; PM3, Initial, symmetry (check, will have to be turned off with large number of atoms), check Elect. Charge (under compute), Total charge (neutral, anion, cation; see individual instructions), Multiplicity (singlet), Print (atomic charges), Converge (click), Global (click).

B.   Save and close each structure/structures when finished.

C.   In your report, there should be a figure caption below each figure (*i.e.* Fig. **1**. This is a …). If there are questions associated with that complex, the answers should be provided in the figure caption.

D.   All structures you generate in Spartan will be copy/pasted into your report with figure captions in numerical order. There should be a maximum of 2 figures per page.

E.   The lab title, your name and date should be on the top of the first page. Below that there should be an index of the figures (1. Water

Molecules, page 2). Figures should not start until after the pre-lab questions are answered!

1. Build water (H_2O) in Spartan and run the structure using the parameters defined above. Once the calculations are complete, save the file under the name "water", copy and paste the water molecule so there are two in your work area and minimize the two structures (see Fig. **1**). The calculation performed assumes these species are in the gas phase. Measure the distance between an oxygen atom on one species and the closest H atom on the other molecule. What type of interaction is this? In addition to copying your 2 water molecules into you report, answer the questions in your figure caption. Be sure to close this file when done.

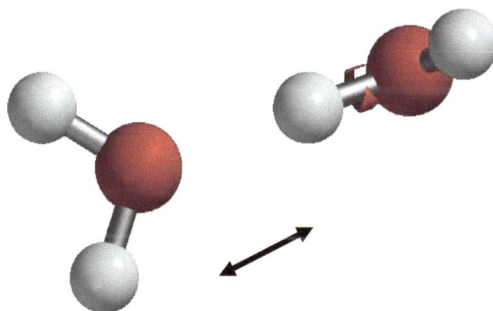

Figure 1: A neutral water molecule is constructed and duplicated. After computational work, the bond distance between the closest oxygen atom on the first molecule and a hydrogen atom on the second molecule is measured. (note atom colors are adjustable).

2. Build a sodium ion (Na^+), (charge = cation) and perform calculations on it. Save this file under the name "sodium". Copy and paste your structure so two ions appear on the screen (see Fig. **2**). Copy/place your two ions in your report. With 2 ions present – do not minimize or perform any calculations on the two atoms - yet. Measure the distance between the two ions, now minimize their energy. What happens? (more than likely both will go off screen and be so far apart, from an atomic perspective it will be difficult to bring both back onto the same

screen). Citing Coulombs Law, explain your observation? Again, answer these questions in the figure caption in your report.

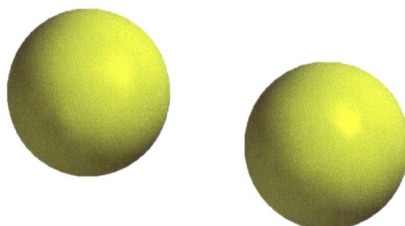

Figure 2: Two Na^+ ions are placed near each before any energy minimization occurs. In this case their distance is 1.257 Å apart.

3. Do the same procedure for Ca^{+2} (charge = dication) that you just did for sodium. Save under the name "calcium".

4. Build a single fluoride (F^-) anion and perform the calculations (charge = anion) and save this structure as "fluoride". Once calculations are complete, copy and paste it so you have two anions next to each other and measure distance (see Fig. **3**). Note the energy box in the lower right hand corner before you minimize the ions. Once you minimize the energy, what happens to the value as the two ions grow further apart? Using Coulombs Law, explain this action.

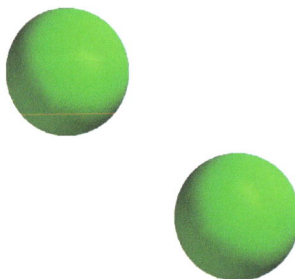

Figure 3: Two fluoride ions before energy minimization takes place.

5. Using Lewis structures, identify the geometry and hybridization of sulfate (SO_4^{-2}). Now build sulfate in Spartan and calculate its parameters (charge = dianion). Save the file under the name "sulfate".

Does this structure have a dipole moment? What are the bond angles? Bond lengths? What are the atomic charges on S? each O? (Remember to copy each structure to your report!)

6. Open your sodium and your fluoride files in different Spartan windows. Copy/paste on into the other window (see Fig. **4**). Minimize the energy and measure the bond distance between the two structures. In terms of Coulombs Law, explain what you observe and compare it to the Na^+ - Na^+ and F^- - F^- interactions.

7. Do the same procedure for sodium and sulfate ($1Na^+$ and $1SO_4^{-2}$). Measure distance from the sodium ion to the nearest oxygen, than from the sodium ion to the sulfur atom – which qualifies as the correct value for the bond distance? Why? Is the Na^+-SO_4^{-2} bond distance the same or different that the Na^+-F^- distance? Explain.

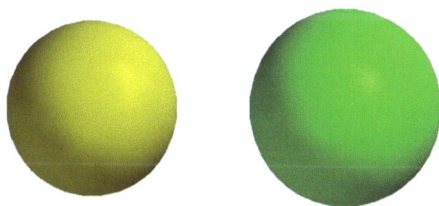

Figure 4: A sodium cation and a fluoride anion are placed next to each other and the energy of the system minimized. Unlike the two cations (Na^+ - Na^+) or anions (F^- F^-), these two ions (Na^+-F^-) are attracted to each other.

8. Do the same procedure for sodium and sulfate, except use two sodium ions ($2Na^+$ and $1SO_4^{-2}$). Is the Na^+-SO_4^{-2} bond distance the same or different that the 1:1 Na^+-SO_4^{-2} distance? Explain. (be sure to copy each workspace system to your document and explain the results in the figure caption).

9. Do the same procedure for calcium and sulfate ($1Ca^{+2}$ and $1SO_4^{-2}$). Is the Na^+-SO_4^{-2} bond distance the same or different that the Ca^{+2}-SO_4^{-2} distance? Explain.

10. Do the same procedure for calcium and fluoride ($1Ca^{+2}$ and $1F^-$). Is the Ca^{+2}-SO_4^{-2} bond distance the same or different than the Ca^{+2}-F^- distance? Explain. (be sure to close flies when completed)

11. Open your "fluoride" file and open "water" file in a separate window. Copy and paste the water into your fluoride window twice (see Fig. **5**). Minimize the ion- molecule system. Measure the distance between the O and H atoms on water and the fluoride ion. Which are the closest? Why? What type of attraction is this classified as (H-bond? Dipole-dipole? Ion-dipole? London force? *etc.*)?

12. Do the same for sulfate and two waters (make same measurements, answer questions, *etc.*) and compare your results to the fluoride-water system completed above.

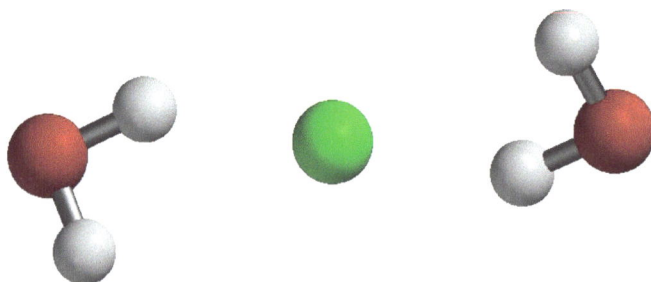

Figure 5: A fluoride anion (green) is attracted to the partial positive charge on a hydrogen atom (white) that is on water.

13. Open your "sodium" file and open "water" file in a separate window. Copy and paste the water into your sodium window twice (see Fig. **6**). Minimize the ion and 2 molecule system. Measure the distance between the O and H atoms on water and the sodium ion. Which are the closest? Why? What type of attraction describes this interaction?

14. Do the same procedure for the calcium (Ca^{+2}) and water and, applying Coulombs Law, which system (Na^+-water, Ca^{+2}-water) produces a stronger interaction? Explain?

Figure 6: A sodium cation is located between two water molecules, which is an example of an ion-dipole interaction. Why are the H's on the two water molecules out of plane with each other?

15. Construct 1-propanol, perform calculations (neutral, singlet) and save it under the file name "propanol". Once calculations are complete, copy, paste and minimize it so you have 2 propanol molecules in your work area (see Fig. **7**). Note how the two molecules, with a polar group (-OH) and a nonpolar component (C_3 chain) align themselves. What is the distance between the closest O and H forming an H-bond between the 2 molecules.

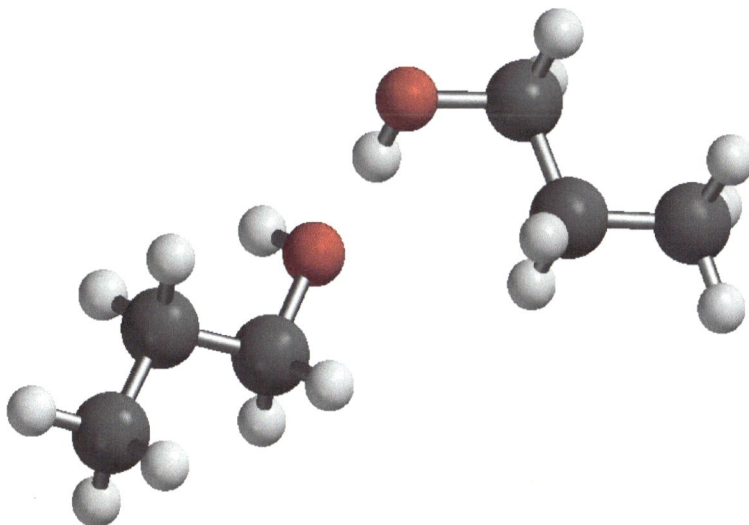

Figure 7: Two 1-propanol molecules are attracted to each other *via* a H-Bond.

16. Construct an octane molecule (C_8H_{18}) in your workspace. Perform calculations on it and save it under the filename "octane". Copy the structure and minimize the energy (Fig. **8**). Measure and record the distance between the five (5) of the closest hydrogen atoms between the two molecules. Compare the distances between the two nonpolar hydrocarbon molecules to that of the distance between two water molecules (above). Does this explain the difference in densities between liquid water (1.0 gram/cm^3) and liquid octane (0.917 g/cm^3)? Is the distance correlated with the difference in surface tension between octane (21.62 mN/m) and water (72.80 mN/m)? Explain.

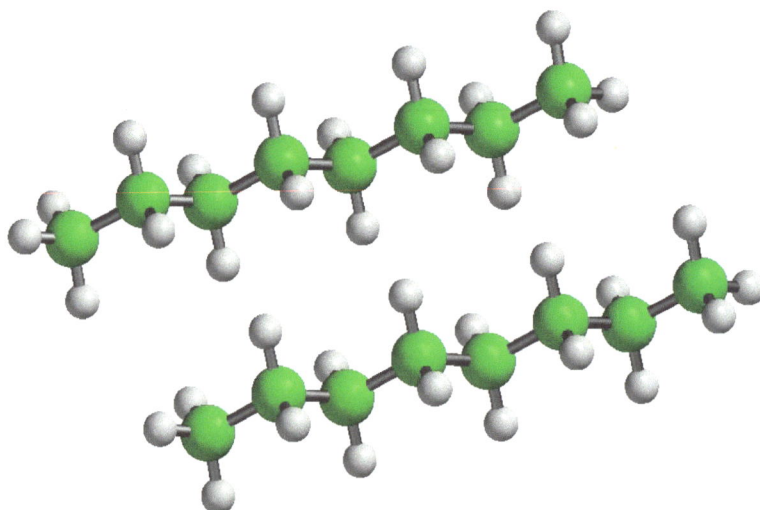

Figure 8: Two octane molecules remain fairly close after energy minimization.

17. You will create a total of three different files/workspaces in this exercise. Be sure to close and save each before moving to the next structure. Set up a table in your report like that in Table **1**. First build a methanol molecule, perform the calculations and save it as "methanol". Copy and paste the structure so you have two of them in your workspace and minimize the system. Measure the H-bond distance between the two molecules (H on one molecule, O on the other) and record it in your table. Close your methanol file and build and calculate an ethanol structure. Copy it and minimize the energy

and measure the H-bond distance between the two structures. With ethanol (and next propanol) measure the distance between the two carbons that are furthest from the oxygen atom – the C's in the methyl groups (see Fig. **9**). Do the same for propanol (make, calculate, copy/paste, measure H-bond distance) and save the file.

Table 1: After performing calculations on methanol, ethanol and propanol systems, record your data in a table with a format like this (in report). Find the vapor pressure and surface tension in the CRC reference book (ask instructor for help with these values).

Species	Distance Between 2 Nearest H,O Atoms	Boiling Point	Density (g/mL)	Vapor Pressure	Surface tension
Methanol		66	0.791		
Ethanol		78	0.789		
Propanol		97	0.804		

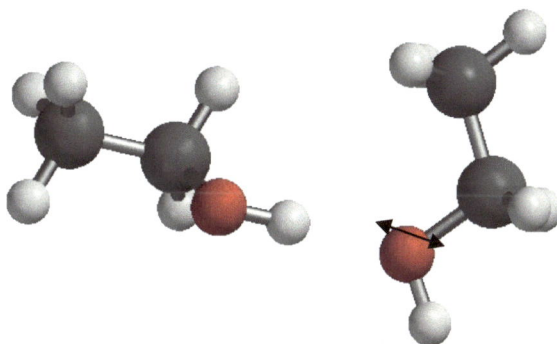

Figure 9: Ethanol molecules show a H-bond between the 2 structures.

18. Open your "water" file in a new work space. (With each aggregate constructed below, minimize it, than perform a Single Point Energy, symmetry (check), Molecular Mechanics (MMFF), neutral, singlet, Converge (check) calculation (do this for all aggregates, water, methanol and ethanol). Molecular mechanics is a lower level of theory than semiempirical, but it can perform the calculations on the larger aggregates using a desktop computer. Note that as the number of water molecules increases, the minimization and calculation time will

also increase. Set up a table that has the columns (Table **1** in your report):

A) # waters (B) molecular volume/Spartan (C) surface area (D) dipole moment (E) image (insert your aggregate in this box) (F) comment (G) volume (calculated from $V=4/3\pi r^3$).

The values in columns B, C, D are found under the icons (click: Display; click, Properties). To obtain the value for column (G), pick two atoms on opposite sides of an aggregate that represents the diameter and measure the distance across the nanodrop and use this value to obtain your radius and your volume. For example, see Fig. **10H**.

Copy/paste your water so there are two molecules in your workspace. Minimize, run calculations, get the needed data for table and save as the file as water_2. What type of bond is holding the two species together (answer question in comment box).

Copy/paste the two molecules to form an aggregate with 4 molecules, minimize, run, get data and save under filename water_4. What shape does the tetra-water aggregate reflect?

Copy/paste the four water molecules so the new aggregate has eight water molecules and minimize it. Again, note the shape of the aggregate defined by the eight oxygen atoms. Measure 3-4 of the angles and distances that define the shape of the aggregate to confirm the shape formed.

Copy/paste the eight molecule aggregate to form a sixteen member aggregate and perform the respective calculations and measurements, save the structures to a memory device. For any of the calculations performed today (water, methanol, ethanol) stop any minimization that takes more than eight minutes (8 min max). Repeat this cycle (1, 2, 4, 8, 16, 32, 64, 128, 256) until you create an aggregate with 256 water molecules. You should have a minimum of five structures per page in your table and save all structures (water_32, water_64, water_128, water_256).

Go through the same procedure for methanol (methanol_2, methanol_4, methanol_8, methanol_16, methanol_32, methanol_64, methanol_128, methanol_256) and ethanol (ethanol_2, ethanol_4, ethanol_8, ethanol_16, ethanol_32, ethanol_64, ethanol_128, ethanol_256) placing the calculated values and images in the table. In the comment box, some concepts that might be briefly discussed: (a) explain the trend or difference observed in the dipole moment of a single molecule of water, methanol or ethanol verses that of an aggregate of molecules (b) are the two volume measurements the same or different? Why? (c) How many H-bonds is a molecule at the surface of a nanodrop involved in compared to a molecule in the middle of an aggregate?

B) When your table is complete, generate five plots in your spreadsheet program (water, methanol, ethanol values on each plot). Each plot should have its own page and its own caption, in numerical order.

a. # molecules (2, 4, 8, 16, 32, 64, 128, 256) on x-axis verses molecular volume (Spartan value) on y-axis. Be sure to designate water, ethanol and methanol as their own series and plot a best fit line through each. Include a graph caption (*i.e.* Graph 1. This graph is.)

b. # molecules (2, 4, 8, 16, 32, 64, 128, 256) on x-axis verses surface area (Spartan value) on y-axis. Be sure to designate water, ethanol and methanol as their own series and plot a best fit line through each.

c. # molecules on x-axis verses calculated volume (from $V=(4/3)\pi r^2$) on y-axis;

d. Calculated volume (x-axis) verses molecular volume (Spartan).

e. Molecular volume (Spartan value) verses surface area (Spartan) on y-axis.

Explain the trends observed in each graph in the caption below that particular graph (example, the slope is $\Delta y/\Delta x$ – what does this value tell us about each solvents aggregates). Does the molecular volume (Spartan value) only include

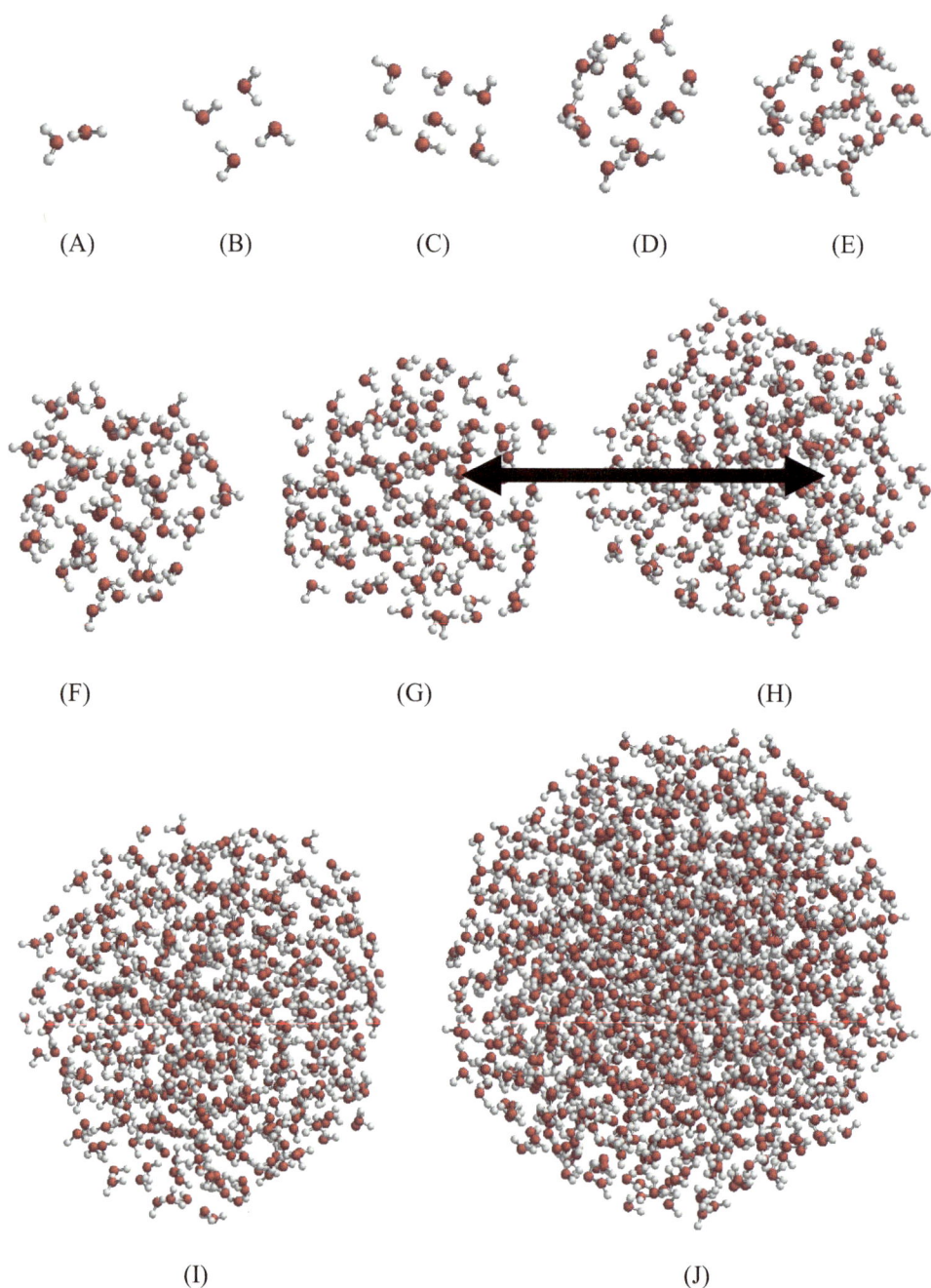

Figure 10: Waters molecules are co-added to form a nanodrop. (A) two (2^1) water molecules linked by a single hydrogen bond (B) four (2^2) water molecules (c) eight (2^3) water molecules (D) sixteen (2^4) water molecules (E) thirty-two (2^5) water molecules (F) sixty-four (2^6) water

molecules (G) one and twenty eight (2^7) er molecules (H) two hundred and fity six (2^8) water molecules (I) five hundred and twelve (2^9) water molecules (J) one thousand and twenty-four (2^{10}) water molecules. (all diagrams are not on same scale). (note minimztion of the largest structures can take hours, your instructor may provide this to you already done).

volume occupied by the water molecules or does it include the space between each species? You can draw this conclusion by comparing the two volumes you calculated for each aggregate – Spartan volume verses $V=4/3\pi r^3$ value. What is the normal density for water at room temperature and pressure? Why is this value different (consider the surface area of the droplet to the spherical mass ratio)? When you minimize these aggregates of molecules, why does it form a sphere and not another structure? (box, triangle, rod, *etc.*) – answer this in the graph captions.

Be sure to save and close your water, methanol and ethanol files.

19. Open your sodium file (*i.e.* Na$^+$) in its own work space and copy/paste it four times (five Na$^+$ ions present). Do not minimize it – yet! Copy/paste it into your report.

    a. Open your water_256, methanol_256, and ethanol_256 files in their own workspaces (you should have four workspaces open; water, methanol, ethanol, sodium ion).

    b. Copy/paste the sodium ions into each of the solvents.

    c. Minimize the energy in the sodium workspace and note what happens to the cations (don't try to find the ions once the minimization is complete). Explain this observation in your sodium figure caption – incorporate Coulombs law into your answer.

    d. Next, minimize the sodium in each solvent (one at a time). When one solvent is complete, zoom into the region where the ions are situated and study the interaction between the ions and the water. Measure three or four of the shortest ion-dipole interactions. Copy and paste the image into your report; crop and zoom in and, using arrows show the ion-solvent interaction. Try not to have more than 10-12 atoms in your cropped image.

e. Do the same for the other two solvents, again noting the distances and copying the image. Save these files as "water_256_Na", "methanol_256_Na" and "ethanol_256_Na". In each figure caption, comment on how your computational observations translate into solubility (*i.e.* is Na^+ soluble in water? Why? Why would Na^+ have a higher solubility in water than ethanol?)

f. Perform the same set of calculations for sulfate in the three solvents that you just performed using the sodium ion (*i.e.* insert SO_4^{-2} in place of sodium). Do the sulfate ions in the workspace behave the same as the sulfate ions in the solvents? Why not?

20. Important Topics to review:

a. Peptide bonds and protein structure

b. Positive (clockwise) and negative (counterclockwise) torsion angles.

c. Ramachandran plot

d. Identifying phi and psi angles in a protein structure.

In this section you will measure the impact that different solvents have on the structure of peptides (and proteins!). Using a web search engine, enter the terms "protein folding, disease" or "protein folding, Alzheimer's" and read about the impact that protein shape can have on a number of diseases (your instructor may ask you to write a 1-2 paragraph overview on this medicinal area).

In Spartan, use the "pep" (peptide) tab and build polyglycine (3 residues). You can adjust two of the torsion angles in the program (click "other" and enter the angles). Enter the following phi, psi angles and generate the tri-peptide (one at a time) but do not minimize the energy:

(0,0), (0,45), (0,90), (0,135), (0,180), (0,-45), (0,-90), (0,-135) (0,-180)

Then generate structures with the following angles:

(45,0), (90,0), (135,0), (180,0),(-45,0), (-90,0), (-135,0), (-180,0)

Once you enter the angle and generate the structure, Do NOT minimize the energy!

Table 2: Generate a table with the following format. Include a side view and a end-on view, coupled with arrows, to illustrate each angle.

Ψ Φ Angles	Φ structure	Ψ structure
(0,0)		(insert Spartan structure, include arrow to identify angle)
(0,45)		
(0,90)		

Maneuver the structure so you are looking down an axis can clearly identify the angle (pick one phi and psi per structure to enter into your table). You can use the "Constrain" icons in Spartan to select four atoms and obtain a phi or a psi angle. Use this approach to identify and confirm the angles entered and help you quickly identify these two important angles that involve four atoms.

Copy/paste each view into your table and use an arrow to identify the angle (Table **2**). This exercise should help you easily identify each angle and also begin to visualize how adjusting the angles impacts the shape of the structure. In building this table you would like to have at least five angles per page (*i.e.* (0,0), (0,45), (0,90), (0,135), (0,180)).

21. Find the structures (peptide sequence) for Methionine-enkephalin and leucine-enkephalin. Write a short paragraph (your words) about the role of each peptide in your report and include their sequences.

22. Develop a new table with the format of Table **3** (you might make this a landscape format):

Table 3: Data for bond angles.

Original phi, psi, angles	Solvent, peptide	Measured phi, psi angles after minimization in nanodrop.	Image of Structure (peptide removed from aggregate, no arrows needed)
+115,-120	Water (256), leucine		
etc.			

23. Build methionine enkephalin and pick a phi angle in the range of -110 to -140 and a psi angle in the range of +110 to +135. Use these values for all enkephalin structures. Do not minimize the structures energy. In a separate window, open you "water_256" nanodrop. Copy/paste you peptide into the drop and minimize the structure. Allow the peptide-water aggregate to minimize for a maximum of three minutes (or until complete). Using the "align" icon, you can move the peptide out of the aggregate and copy/paste it into its own window. Measure three phi and psi angles in this structure. Close the aggregate (be sure to have water_256 saved separately)

24. Build methionine enkephalin and use the same phi and psi angles described above. Do not minimize the structures energy – yet! In a separate window, open you "methanol_256" nanodrop. Copy/paste you peptide into the drop and minimize the aggregate. Allow the peptide-methanol aggregate to minimize for a maximum of three minutes (or until complete). Remove the peptide from the aggregate and copy/paste it into its own workspace. Measure and record the values of the three psi angles in your structure. Copy this structure into your table.

25. Build methionine enkephalin and enter the same phi and psi angles described above. Do not minimize the structures energy – yet! In a separate window, open you "ethanol_256" nanodrop. Copy/paste you peptide into the drop and minimize the aggregate. Allow the peptide-ethanol aggregate to minimize for a maximum of three minutes (or until done). Remove the peptide from the aggregate and copy/paste it

into its own workspace. Measure and record the three psi and phi angles in your structure.

26. Build leucine enkephalin and enter the same phi and psi angles described above. Do not minimize the structures energy – yet! In a separate window, open you "water_256" nanodrop. Copy/paste you peptide into the drop and minimize the aggregate. Allow the peptide-water aggregate to minimize for a maximum of three minutes (or until done). Remove the peptide from the aggregate and copy/paste it into its own workspace. Measure and record the three psi and phi angles in your structure.

27. Build leucine enkephalin and enter the same phi and psi angles described above. Do not minimize the structures energy – yet! In a separate window, open you "methanol_256" nanodrop. Copy/paste you peptide into the drop and minimize the aggregate. Allow the peptide-methanol aggregate to minimize for a maximum of three minutes (or until done). Remove the peptide from the aggregate and copy/paste it into its own workspace. Measure and record the three psi and phi angles in your structure.

28. Build leucine enkephalin and enter the same phi and psi angles described above. Do not minimize the structures energy – yet! In a separate window, open you "ethanol_256" nanodrop. Copy/paste you peptide into the drop and minimize the aggregate. Allow the peptide-ethanol aggregate to minimize for a maximum of three minutes (or until done). Remove the peptide from the aggregate and copy/paste it into its own workspace. Measure and record the three psi and phi angles in your structure.

29. Build methionine enkephalin in a workspace with no solvent, use the starting angles you used before and minimize this structure with no solvent present. Measure the phi and psi angles.

30. Build leucine enkephalin in a workspace with no solvent, use the starting angles you used before and minimize this structure with no solvent present. Measure the phi and psi angles.

31. Generate two Ramachandran plots using your data. One for the methionine enkephalin data and one for the four sets of leucine data. Be sure that each axis is from -180 to +180 and there are four quadrants in your plot. Each plot should have four data sets (no solvent, water, methanol, and ethanol). Each data set should have its own unique symbol (water = box, ethanol=cross, *etc.*) that is indicated in a legend box.

32. Once your plots are complete, some questions will be answered (separately) below the plot:

- Do the final phi, psi values indicate a beta or alpha sheet structure? If an alpha sheet, is it right or left handed?

- Are there structural differences within MenK and LenK as a function of solvent?

- Using the predefined semiempirical calculation procedure, calculate the dipole moment (D) and molecular volume (V) for water, ethanol and methanol. Than calculate a D/V ratio (dipole moment/molecular volume) for each molecule.

- Plot the D/V ratio for each solvent (y-axis) *vs.* the calculated molecular volume of the peptides (both leucine and methionine). This graph should have six points. Based on the structure and its subcomponents (polar, nonpolar groups) develop and argument about how different solvents will impact the energy involved in folding a peptide or a protein.

Some key Reminders in Report Preparation

1. Title (Chemistry in a Nanodrop, 18 point font), name date: centered on top of page one.

2. Index of all figures, graphs, tables starting on the first page below title. This should be single spaced. Figures first (Fig. **1**. Water Molecules, p. 3), than a graph section and a table section.

3. Every figure, graph, table should have its own caption, written in full sentences, that explains the image. (*i.e.* Fig. **1**. Two water molecules are attracted to each other by a hydrogen bond and are 1.77 Angstroms apart.)

4. Figs. **1,2,3,4**.; graphs 1,2,3.; Tables **1,2,3**. each have their own numbers.

5. Center figures on the page. Typically two figures per page is desired.

6. Figures from Spartan should have a white background.

7. Crop and expand figures as needed. Use arrows to indicate interactions or structures discussed in the caption.

8. Start tables and graphs on a new page. One graph per page.

9. Equations on its own line with a number. $PV = nRT$ (1)

10. Identify variables and units used in equations. P is the pressure and is in atmospheres (atm's).

11. Label graph axis with unit. Time (sec)

12. When using a best fit ($y = mx + b$), include the equation and correlation coefficient on the graph.

13. Your name and title on the top of each graph (*i.e.* Velocity of Car Data, Joe Smith)

14. Margins on each page should be 1x1x1x1 inch.

15. 11 or 12 point font, New Times Roman or Arial. Page numbers, lower right hand corner.

16. Figure, graph, table captions should be single spaced. Other parts of the lab should be double spaced. Staple your report! No cover sheets or folders.

REFERENCES

[1]   Lenz, A.; Ojamäe, L. Computational studies of the stability of the $(H_2O)100$ nanodrop, *Theochem.-J Mol. Struc.*, **2010**, *944,* 163-167.

[2]   Müller, A.; Bögge, H.; Diemann, E. Structure of a cavity-encapsulated nanodrop of water, *Inorg. Chem. Commun.*, **2003**, *6*, 329.

[3]   Bielański, A.; Małecki, A.; Lubańska, A.; Diemann, E.; Bögge, H.; Müller, A. The behavior of in a water nanodrop encapsulated within a highly charged porous metal–oxide nanocontainer: A thermoanalytical study, *Inorg. Chem. Commun.*, **2008**, *11*, 110-113.

Send Orders of Reprints at reprints@benthamscience.net
Computer Based Projects for a Chemistry Curriculum, 2013, 183-185   **183**

CHAPTER 17

Atoms in Space: Isomers, Coordination Compounds and Other Structures

Thomas J. Manning[*] and Aurora P. Gramatges

Department of Chemistry, Valdosta State University, Valdosta, Georgia, USA, and Instituto Superior de Tecnología y Ciencias Aplicadas, La Habana, Cuba

Abstract: In this exercise students will use the molecular modelling software to build and study different types of isomers. Students will use the molecular modelling software to build and study different coordination compounds. Working with dynamic structures in 3D allows students to visualize and understand geometric factors that are difficult to comprehend with a flat (2D) image.

Keywords: Isomers, coordination compounds, molecular modelling, semiempirical calculations.

INTRODUCTION

Isomers are molecules with the same empirical formula but with different atomic arrangements in space. Isomers of the same compound can have different chemical and physical properties. The basic types of isomers are shown in Table **1**. In this exercise you will be given different structures to build in two and three dimensions. For the 2D structures use a program such as ISIS or WordArt to build your structures. Molecular mechanics calculation is the most commonly used type of calculation in inorganic chemistry [1-3]. Using the modelling software you will also run semiempirical (PM3) calculations on the structures built in 3D (Spartan).

---

*Address correspondence to Thomas J. Manning: Department of Chemistry, Valdosta State University Valdosta GA 31698, USA; Tel: 229-333-7178; E-mail: tmanning@valdosta.edu

Table 1: Create a table similar to this for your report. Because your structures may consume space, this might be in a landscape format. There are coordination compounds and isomers in this table. Only assign a central atom, coordination number and charge on the central atom for coordination compounds. Import your 2d and 3D (Spartan) structures to the table. In the column "unique aspect" briefly define the geometry or describe what makes it unique from other geometries.

Coordination couple	Name	Central Atom	2D structure	3D structure	Unique aspect	Coordination number	Charge of central atom
$K_3[Fe(CN_6)]$ (coordination; single structure)							
$[N(H_2O)_6]Cl_3$ (coordination; single structure)							
$Na_2[Cu(H_2O)_2Cl_2\,Br_2]$ (coordination; 2 isomers)							
$[Zn(en)(CO_3)O]$ (coordination; 2 isomers)							
$Na[Cr(SO_4)(OH)_2(H_2O)_2]$ (coordination; 2 isomers)							
$Na_3[Fe(ox)_3]$ (coordination, 2 isomers)							
mer-$[CoCl_3(NH_3)_3]$ (coordination, single structure)							
fac-$[CoCl_3(NH_3)_3]$ (coordination, single structure)							
cis-$[CoCl_2(NH_3)_4]^+$ (coordination, single structure)							
trans-$[CoCl_2(NH_3)_4]^+$ (coordination; single structure)							
$[Co(EDTA)]^-$ (coordination, single structure)							
Cis-Plat (coordination, single structure). Famous cancer drug							
Trans-plat (coordination, single) little cancer activity.							

D and L forms of leucine (which is + and - ?)								
Use two structures of Bromochlorofluoromethane to demonstrate enantiomers.								
Use three structures of C_5H_{12} to demonstrate structural isomers.								
Use two examples of rotamers using $C_2H_4Cl_2$								
Two structure of butane to demonstrate gauche and anti-conformations.								
An example of a linkage isomer using SCN⁻ (create two structure with SCN attached different)								
Use two polyglycine rings (12 residues each) to demonstrate a Catenane								
Use a strain of DNA to make a molecular knot. (Topoisomer)								
R and S forms of serine								
Demonstrate a rotaxane structure using α-cyclodextrin as the macromolecule and 1,1,10,10-tetraphenyldecane as the dumb bell molecule.								

REFERENCES

[1]   Hay, B. P. Methods for molecular mechanics modelling of coordination compounds, *Coordin. Chem. Rev.*, **1993**, *126*, 177-236.

[2]   Deeth, R. J.; Anastasi, A.; Diedrich, C.; Randell, K. Molecular modelling for transition metal complexes: Dealing with d-electron effects, *Coordin. Chem. Rev.*, **2009**, *253,* 795-816.

[3]   Nicolau, D. V.; Yoshikawa, S. Molecular modelling of Me2+- (8-hydroxy-quinolinate)2 complexes using ZINDO and ESSF methods, J. Mol. Graph. Model., **1998**, *16*, 83-96.

CHAPTER 18

Constructing and Visualizing Some Common Materials

Thomas J. Manning* and Aurora P. Gramatges

Department of Chemistry, Valdosta State University, Valdosta, Georgia, USA, and Instituto Superior de Tecnología y Ciencias Aplicadas, La Habana, Cuba

Abstract: In this exercise, students will learn how to use a modeling program to build a lattice structure. It will teach students how to construct a sheeted material such as graphite, how to construct an intercalated compound, how to construct a fullerene (buckyball, C_{60}) and a aza-fullerene ($C_{48}N_{12}$), and how to construct a carbon nanotube from chains of carbon molecules. It aims to improve a student's ability to visualize 3D structures.

Keywords: Molecular modelling, materials, intercalation, fullerene, nanotube, allotropes.

INTRODUCTION

Molecular modelling software serves as an excellent visualization tool for three dimensional structures. This lab exercise is focused on having students build a series of materials including a structure, graphite, a graphite intercalated compound, diamond, a doped diamond structure, carbon nanotubes, C_{60} and $C_{48}N_{12}$, a boron nitride sheet and a boron nitride nanotube [1-3].

Pre-Lab Questions

1. Draw and briefly describe a Simple Cubic (SC), body centered cubic (BCC), face centered cubic (FCC), primitive FCC, primitive hexagonal, and cubic closest packed.

2. What is a unit cell? Why does a unit cell include fractions (1/8, ¼, ½) of atoms in its corners and sides? Describe S1, S2 and S3 unit cells with 2D drawings.

Address correspondence to Thomas J. Manning: Department of Chemistry, Valdosta State University Valdosta GA 31698, USA; Tel: 229-333-7178; E-mail: tmanning@valdosta.edu

3. What is Coulombs Law (equation, variables, units, what it models)

4. There are three allotropes of carbon; fullerenes, graphite and diamond. Briefly describe the structures (hybridizations, bond angles, bond lengths, *etc.*) for each of these. Also, describe the geometry of a single walled carbon nanotube (SWNT).

5. What is a graphite intercalated compound (GIC)? Give two applications of GIC's in commercial or research applications. Also, briefly define what a van der Waals force is and how it applies to graphite sheets.

Exercise 1: Sodium Chloride

Students are taught that simple salts are held together by electrostatic forces, interactions that are described by Coulombs Law. The parameters of these structures are dictated by ionic radius and charges. This exercise assumes the student has some experience in working with Spartan. The computational parameters that will be used are Single Point Energy; Semiempirical (PM3), Start (Initial), Check Symmetry, Check electrical charges, total charge (cation or anion), Multiplicity (singlet).

1. First we will build a sodium cation. Select Na from the "Exp" tab and remove any bonds so it appears as a single sphere (Fig. **1A**). Perform the calculations outlined above but select "cation" for charge and singlet for multiplicity. In a separate workspace construct a chloride ion (Cl⁻) and be sure to select "anion" for charge. Copy and paste one ion into the workspace of the other, minimize the system so a NaCl (Fig. **1B**) forms.

Begin to copy and paste the NaCl structures unto itself (B). You might have to move one or two of the ions so repulsion forces don't push ions off the workspace. After each duplication, minimize the lattice. Create a table with 2 columns. The first column is labeled "Description" and the second column is labeled "Structure". You should have aggregates in your report for 2, 4, 8, 16, 32,

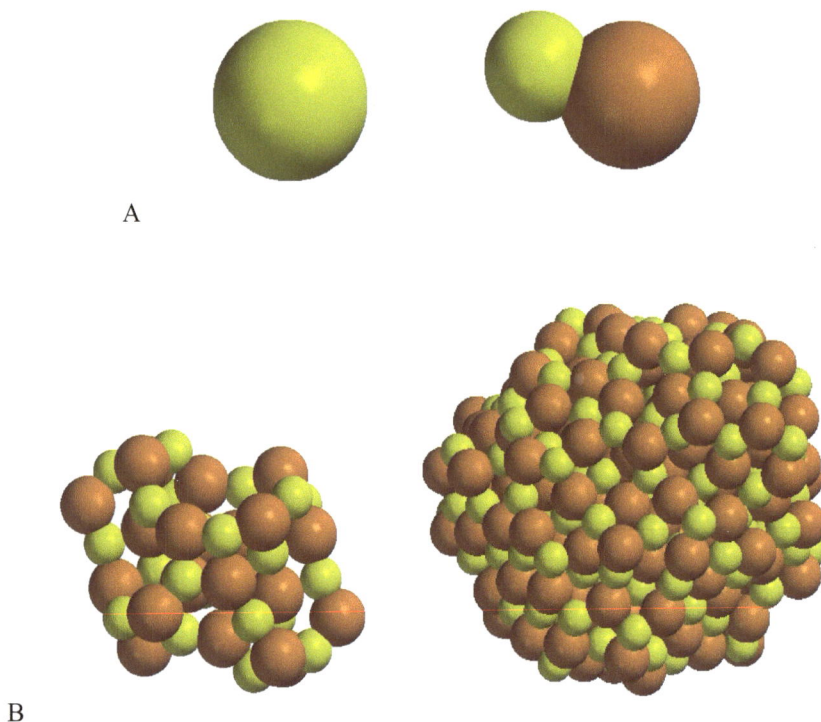

Figure 1: A sodium ion (Na^+) appears as a sphere in Spartan.

64, 128, and 256 NaCl pairs. In the first column provide the number of NaCl pairs present (*i.e.* 2, 4, 8, 16, 32,….256), provide up to five Na-Cl distances for ions touching each other and estimate the volume and surface area. Perform semi empirical (PM3, neutral) or molecular mechanics calculations to get these values. As the lattice structure grows (see Fig. **2A** and **2B**), a structure should begin to emerge. Once you construct and measure the largest aggregate or salt structure (256 NaCl pairs), identify the unit cell and its dimensions. Sodium chloride (halite) crystal structure has six neighbor ions that are in the inner sphere and possess an octahedral geometry (called *cubic close packed* (ccp)). The NaCl halite structure can be visualized as a FCC lattice of chloride ions, with the cations occupying holes. Does your lattice have this geometry?

The three dimensional packing efficiency (PE) of a lattice can be calculated by:

$$P.E. = (\text{volume of spheres}) / (\text{volume of cell})$$

Where $V = 4/3\pi\, r^3$. Estimate the PE of your unit cell and record your value in the column/box with the largest (256) structure.

Once this is complete, construct a second table with the same format as the first. It will be used for your CsCl data. In different workspaces construct Cs^+ and Cl^- species and, after performing calculations, move the ions into the same workspace. Construct, calculate and measure the same parameters as above for the successive structures (2, 4, 8, 16, 32, 64, 128, 256) of CsCl. Analyze your structure and unit cells and determine if this is a simple cubic structure.

Exercise II: Carbon Structures

The two best-known allotropes of carbon are diamond and graphite. They differ in their physical and chemical properties because of differences in the arrangement and bonding of the atoms. In graphite structure, each carbon atom is sp^2 hybridized and in diamond atoms are sp^3 hybridized. In graphite the sheets of carbon atoms are six member rings. If an atom or molecule is placed between the sheets of carbon, it is known as a graphite intercalated compound (GIC). The discovery of fullerenes was first reported in 1985 by Harold W. Kroto, Robert F. Curl, and Richard E. Smalley. The most common fullerene obtained is composed of 60 aromatic carbon atoms, C_{60}. $C_{60,}$ is a closed cage structure that can be constructed by properly connecting twelve cyclopentadiene structures. The nomenclature system for pure carbon fullerenes is based on a numerical code. C_{60} $(56^5(56)^5(65)^56^55)$. This first ring is a five (5) member ring (see Fig. **2A**), the second level is five rings each with six (hexagons) member rings (Fig. **2B**), followed by 5 and 6 member rings, five times each (Fig **2C**). The architecture of C_{70} is described as $56^5(56)^56^5(65)^56^55$. Using Fig. **2** as a reference, construct a C_{30} structure. It is described as a Bucky-bowl and is ½ of a buckyball. Once your buckybowl is complete it will be copied, pasted and connected to itself to form a symmetric C_{60}. Copy this spherical allotrope of carbon into your report and using the normal computational parameters (single point energy, semiempirical, PM3, neutral, *etc.*) calculate the surface area, volume and dipole moment. Also measure and record ten carbon-carbon bonds in the structure (are they the same? Different? Explain what this means in terms of delocalization?).

(A)

(B)

(C)

(D)

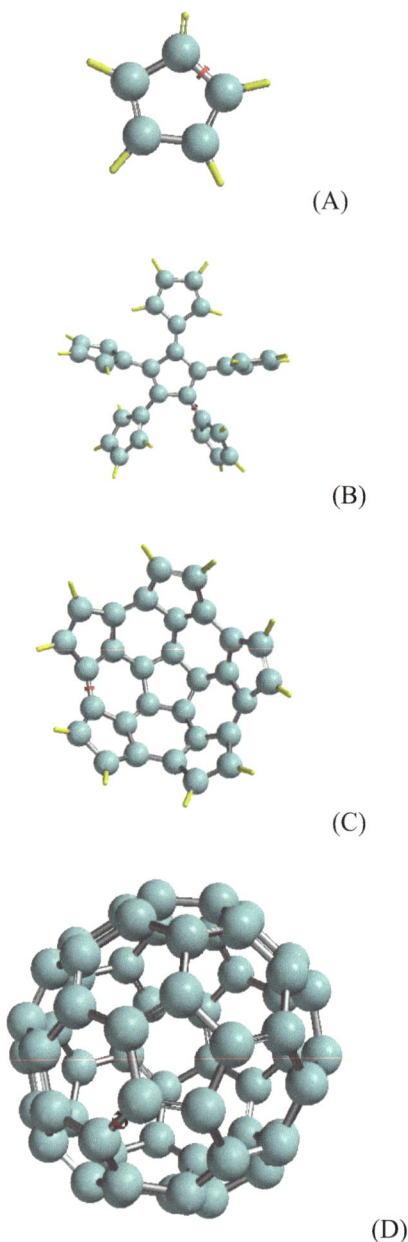

Figure 2: (A). Construct a five member carbon ring with sp^2 hybridized carbon. There should be two double bonds within the structure and a third protruding from the structure. (B) The ring is copied and pasted five times in the same workspace resulting in six 5-membered rings in the workspace. The connection sequence of the five rings to a central ring is critical. When connecting the rings, the protruding double bond should be in the ortho position. (C) When connecting the rings to the central ring, connect the double bonds first. As you connect these bonds, six member rings will form. Ten

single bonds should be protruding from the bucky-bowl (C_{30}). (D). Copy and paste your C_{30} bowl structure into the same window and connect bonds so that only six member rings form.

Fig. **3** provides visual details for the construction of a carbon nanotube (10,0). Construct a (10, 0) tube that is at least 5 nm long. Once it is complete, copy the structure to your report and answer the following questions associated with the structure (A) What is its volume? (B) Can you cap the end of it with a C_6 (benzene type) structure? Provide a visual (structural) evidence for why or why not this structure will function as a cap.

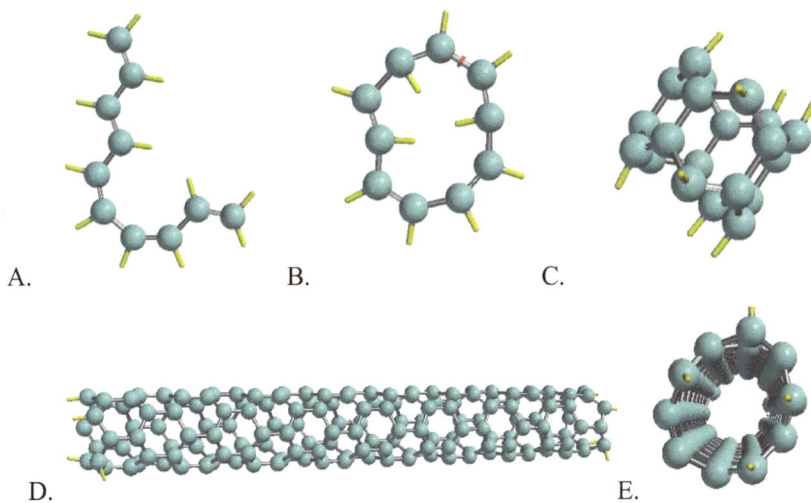

Figure 3: (A). Using sp^2 hybridized carbon atoms, construct a carbon chain with ten atoms. All of the double bonds should be in the chain and only single bonds protrude from the chain (B) Connect the two ends of the chain to form a ring (hence the (10,0) SWNT name). (C) Copy and paste the ring in the same workspace and connect every other bond so six-member rings form. Every other atom should have a bond protruding. (D) Copy and paste this unit, connecting every bond to form a mini-tube with 4 rings, than copy and paste to form a mini-tube with 8 rings, *etc.* (E) When you turn the structure and look down the tube it should be highly symmetric.

Your instructor may now assign you additional fullerene or nanotube structures to construct such as: (A) Unsymmetrical C_{60} ($5^5 7^5 5^{10} 7^5 5^5 5$) (B) C_{70} ($56^5 (56)^5 6^5 (65)^5 6^5 5$) (C) two isomers of C_{96}: C_{96} ($65^6 7^6 (56)^6 (65)^6 7^6 5^6 6$) and C_{96} ($66^6 (56)^6 (66)^6 (65)^6 6^6 6$) (D) nanotube (16,0). (E) Spherical C_{60} with two helium atoms trapped inside. Calculate the dipole moment of the endohedral structure (F) Spherical C_{60} with 38 hydrogen atoms covalently bound outside the structure. Calculate the dipole moment of the exohedral structure.

You will now construct a sheet of graphite. Graphite can be rolled up in an unsymmetrical fashion to form nanotubes with (6,6) and (9,10). The nanotube system described above for nanotubes such as (10,0), (12,0), (14,0), or (20,0) is typically easier to construct in a molecular modelling program. Study Fig. **4** closely

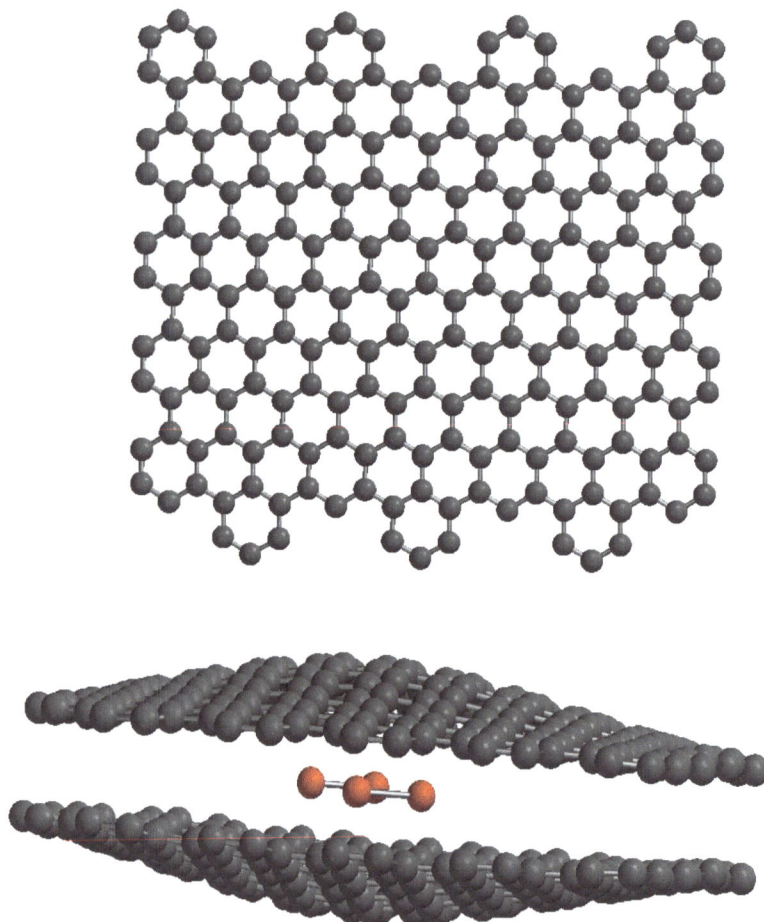

Figure 4: (A) A graphite sheet is composed of sp^2 hybridized carbons and six sided rings. (B) A graphite intercalated compound with 2 bromine molecules sandwiched between the two sheets.

and determine a strategy for constructing the sheeted structure that is composed entirely of sp^2 hybridized carbon. Construct a sheet that is approximately square (length ≈ height) and is at least 9 nm^2. Look for a repeating unit that allows you to copy and paste a section of graphite to make a larger sheet as opposed to adding one carbon atom at a time. Once this is copied to your report, calculate the number of carbon atoms/nm^2. Next copy and paste your structure into the same workspace so you have two sheets. After your minimize it, measure at least ten distances between the two sheets. In a separate workspace, construct a Br_2 molecule and copy two Br_2's between the double graphite sheet areas and minimize the structure. This is a simple representation of a graphite intercalated compound. Using at least ten points, measure the average distance between the two sheets. Again, copy it to your report and provide a brief description.

The next structure to build is a carbon crystal or diamond. It is composed of carbon atoms that are all sp^3 hybridized. As shown in Fig. **5**, build a nano-crystal of diamond that consists of at least 200 atoms with the length, width and height of the structure all with similar dimensions. Copy the structure to your report and calculate its volume. Also, with the bare bonds at the surface of the crystal, what is likely to happen to this under normal, atmospheric conditions?

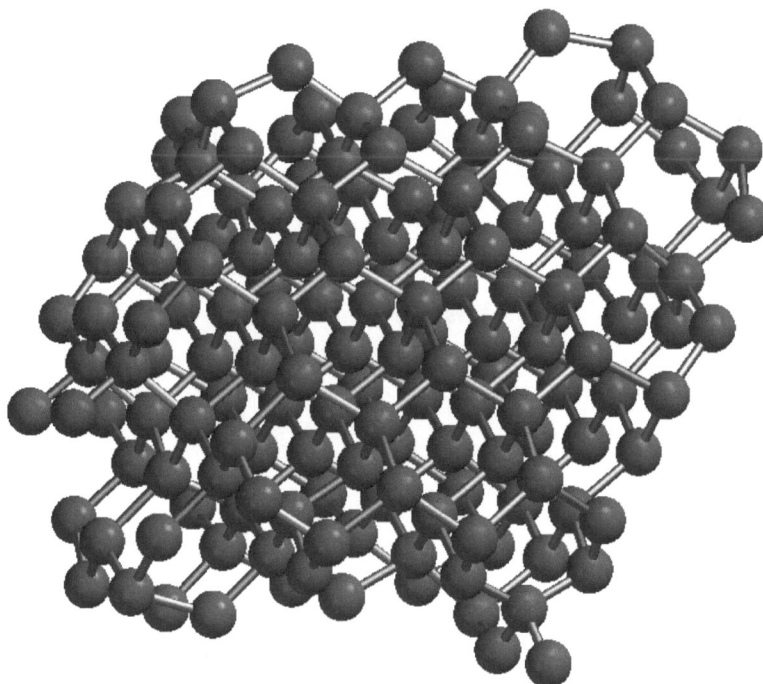

Figure 5: A diamond structure is composed of all sp^3 hybridized carbons.

The final structure to build is the aza-fullerene, $C_{48}N_{12}$. This will be composed of twelve C_4N_1 units and is capable of producing an extraordinarily large number of isomers. Fig. **6** outlines the construction of a single isomer of the aza-fullerene $C_{48}N_{12}$. The first five member ring consists of four sp^2 hybridized carbons and one sp^3 hybridized nitrogen atoms. All of the double bonds should be contained within the ring and only five single bonds are protruding from the structure. There is a central C_4N_1 unit surrounded by 5 identical units (Fig. **6B**). Unlike the construction of the C_{30} which required the builder to account for double bonds, the construction of the $C_{24}N_6$ aza-buckyball only has single bonds giving rise to more flexibility in the construction. Finally, once an aza-buckybowl is constructed, copy and paste it into the same workspace and connect it to form the spherical structure. While this construction focuses on using pyrrole as a building block, an aza-fullerene can be, hypothetically from $C_{59}N_1$ to C_1N_{59}. Within most of these empirical formulas, there exist many potential isomers (at least we can model them – the complete synthesis of an aza-fullerene has never been reported in the literature (although work in this lab has suggested the formation of a hydrogenated aza-fullerene in a high voltage discharge).

There also is no simple method of naming or identifying aza-fullerenes currently available so we developed one for this exercise. C_{60} only has carbon atoms so structures can be identified by the order and size of rings. Here we outline a numerical method to distinguish isomers of $C_{48}N_{12}$ constructed with a C_4N_1. It starts with a central ring (Fig. **6A**) that is made from sp^2 hybridized carbons and a sp^3 hybridized nitrogen. This structure is than copied five times and each one is connected to the central pyrrole. Since the bonds protruding only have single bonds, these structures can be arranged a number of ways. So, even if the ring sizes (*i.e.* 5 member ring) are the same, the physical properties, such as volume, surface area and dipole moment, are different (see Table **1**, Fig. **8**) due to the placement of the nitrogen's. This led us to develop a numerical system to describe each structure. It starts with the central C_4N_1 unit, which is numbered from 1-5 (Fig. **7A**). The first ring of C_4N_1 units is also numbered from 1-5 BUT the #1 atom is the atom attached to the central pyrrole ring. Then, moving in a clockwise direction, the superscript is the position that nitrogen is in (1-5).

This system lets us easily name aza-bucky-bowls. Two aza-buckybowls can be joined to form an aza-fullerene ($C_{48}N_{12}$). Naming of the bucky-balls builds on the bowl numerical system. It appears as $1^3 2^2 3^1 4^2 5^1 - 72°$ - $1^3 2^2 3^1 4^2 5^1$. The first sequence gives the order of the bowl in the back, and the second sequence is the order of the bowl in front. 72° (or 1/5 of 360°) is how many degrees from the nitrogen atom in the central ring is rotated clockwise from the nitrogen in the back ring (see Fig. **8**).

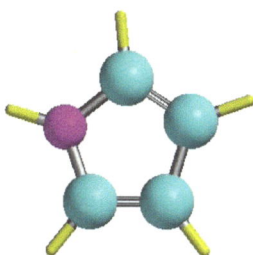

A.

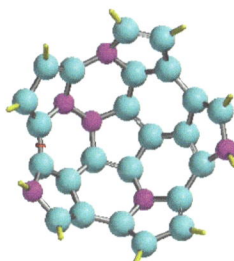

B.

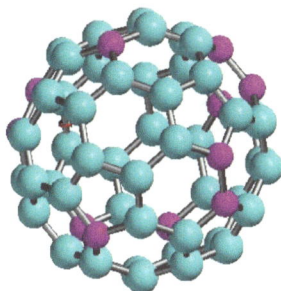

C.

Figure 6: (A) A C_4N_1 subunit is the building block of $C_{48}N_{12}$. All of the double bonds are held within the five member ring (B). Five of the rings are attached to the central C_4N_1 unit but depending on the attachment arrangement, it produces different isomers. (C) The buckybowl in (B) is copy, pasted, and connected to form a spherical allotrope of carbon.

A.

B.   C.

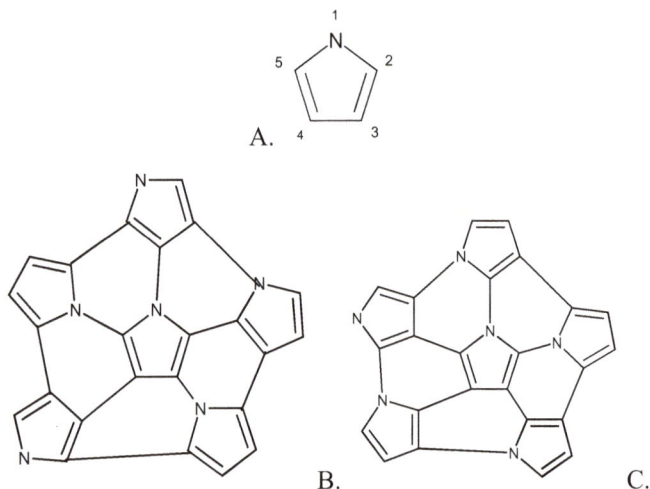

Figure 7: (A). The central pyrrole unit is numbered from 1-5 starting with the nitrogen atom. (B) The five pyrroles that are attached to the central pyrrole are numbered from 1-5 with #1 being the atom attached to the central pyrrole and the superscript represents the position of the nitrogen The nomenclature system developed here would give the name of "$1^3 2^2 3^1 4^3 5^1$" for this structure. (C). The nomenclature system developed here would give the name of $1^4 2^1 3^4 4^4 5^2$.

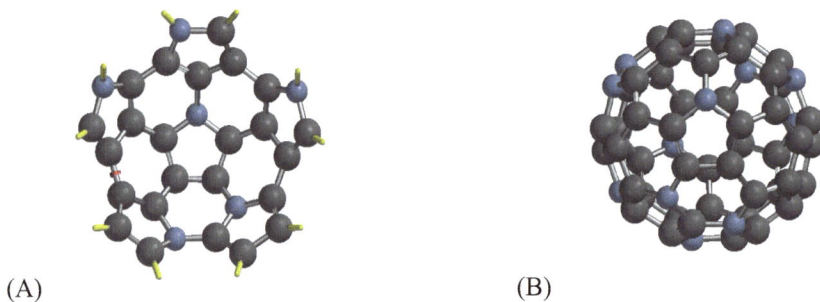

(A)   (B)

Figure 8: (A) The buckybowl $1^3 2^3 3^1 4^2 5^4$. This structure is than copied and the two bowls are joined to form a bowl. When attaching the two identical halves a number of isomers can again form. The numerical representation for this geometry is $1^3 2^3 3^1 4^2 5^4 - 256° - 1^3 2^3 3^1 4^2 5^4$. The 256° indicates how much the nitrogen on the back ring is rotated from the nitrogen on the front ring. Using the "label" command in Spartan, you can keep track of the atom numbers (there are 12 five member rings).

In a 2D program (*i.e.* ISIS) make the following bucky-bowl's (a) $1^2 2^5 3^1 4^3 5^2$ (b) $1^3 2^3 3^2 4^2 5^4$. Once this is complete than make each in your molecular modelling program (Spartan). Once these four structures (two in 2D, 2 in 3D) are complete and copied into your report. Connect the two bowls into the same workspace and make five different structures by varying the bonding order of the bowls. In addition to naming them using the numerical method described above, calculate their dipole moment, molecular volume and surface area and place these values, along with their name and a copy of the structure in a five column table (name, structure, dipole moment, molecular volume, surface area are the headers). In each structure be sure to have the $1^2 2^5 3^1 4^3 5^2$ bowl in the front with the nitrogen on the central $C_4 N_1$ unit at 12 o'clock.

REFERENCES

[1]   Man, Z.; Pan, Z.; Xie, J.; Ho, Y. Molecular dynamics simulations of the C70–graphite interaction, *Nucl. Instrum. Meth. B*, **1998**, *135*, 342-345.

[2]   Man, Z.Y.; Pan, Z.Y.; Ho, Y.K. The rebounding of C60 on graphite surface: a molecular dynamics simulation, *Phys. Lett. A*, **1995**, *209*, 53-56.

[3]   C.R. Wood, N.T. Skipper, M.J. Gillan, Ca-intercalated graphite as a hydrogen storage material: Stability against decomposition into CaH$_2$ and graphite, *J. Solid State Chem.*, **2011**, *184*, 1561-1565.

Send Orders of Reprints at reprints@benthamscience.net

CHAPTER 19

Supercritical Fluid of Carbon Dioxide and Carbon Nanotubes

Thomas J. Manning[*] and Aurora P. Gramatges

Department of Chemistry, Valdosta State University, Valdosta, Georgia, USA, and Instituto Superior de Tecnología y Ciencias Aplicadas, La Habana, Cuba

Abstract: In this exercise students will learn to use a carbon nanotube as gas phase test tube, to study the transition from a gas to a supercritical fluid for CO_2. They will also learn to compare the Ideal Gas Law to the van der Waals equation.

Keywords: carbon dioxide, supercritical fluid, carbon nanotubes, Ideal Gas Law, van der Waals equation

INTRODUCTION

Carbon dioxide is a molecule that is critical to a number of natural and manmade processes. These range from its role as a basic food source in plants to its controversial role as a greenhouse gas. Its role as a product of combustion reactions is well known,

$$CH_4(g) + 2O_2(g) \rightarrow CO_2(g) + 2H_2O(l)$$

Carbon dioxide is fairly easy to make into a supercritical fluid because its T_c and P_c are readily achieved with common equipment [1, 2]. It is the primary species used to extract caffeine in decaffeination processes [3, 4].

Pre-Lab Questions

1. Write a balanced reaction for the combustion of octane, coal and charcoal.

2. What is a supercritical fluid? Using a 2D drawing program, draw the phase diagram of CO_2 and H_2O. Include/label the T_c and P_c (critical temperature and pressure) and the triple point for both species on your graphs.

---

*Address correspondence to Thomas J. Manning:** Department of Chemistry, Valdosta State University Valdosta GA 31698, USA; Tel: 229-333-7178; E-mail: tmanning@valdosta.edu

3. What is the Ideal Gas Law? The van der Waals equations? Identify the parameters and their units in each equation. Under what conditions do you use each?

4. What are the specific parameters "a" and "b" and their units for CO_2 used in the van der Waals equation?

Exercise

You will use a capped nanotube in this computational exercise. You may construct a nanotube or your instructor will give you a nanotube Spartan file. If you construct one, you can use structures made in Exercise 18.

Specifically, the nanotube caps are constructed from buckyballs. Look at your bowl and estimate the type or diameter of nanotube (*i.e.* (10,0), (20,0), *etc.*) that can be capped by ½ of a buckyball. Your (10,0) tube will be too small and a tube like (50,0) would be too big. Your tube should be approximately the same diameter and length as that shown in Fig. **1**.

In a separate workspace create a CO_2 molecule (use EXP tab to get =C=), run calculations (single point energy, semi empirical, PM3, neutral, singlet) on the structure and save the file (your structure should look like Fig. **2**). Copy and paste one CO_2 molecule into your nanotube and minimize the energy. Save the file under the name "nanotube_CO2_1". Copy/paste this structure into your report and answer the questions in its figure caption (show work, with units, for calculations). For all calculations in this exercise, assume the supercritical temperature (you'll calculate the pressure). Calculate the volume, in nm^3, inside the nanotube (hint: measure its height and inside diameter). Using PV=nRT, calculate the pressure inside the tube with a single CO_2 molecule (hint, 1 molecule/(6.023×10^{23} molecules CO_2/mol) = moles of CO_2). Also, use the van der Waals equation and calculate the pressure for each pressure.

Paste a second CO_2 molecule inside the nanotube and minimize the energy. Measure the distance between the two molecules. Using PV=nRT and VDW's, calculate the pressure inside the tube with two CO_2 molecules. For all structures, copy/paste the molecule into your report and answer the questions below in the figure caption. Paste a third CO_2 molecule inside the nanotube and minimize the energy. Measure the distance between the three molecules (*i.e.* 1→2, 2 → 3) and average the value. Using PV = nRT and VDW's, calculate the pressure inside the tube with three CO_2 molecules. Keep adding one CO_2 molecule at a time until the pressure inside the tube is at least three times the value of the critical pressure. After each addition, calculate the pressure and record the distances between each adjacent molecule and average the value. Once you've achieved triple the critical pressure, plot the average distance between the adjacent CO_2 molecules verses the calculated pressure (Fig. **3**). You should have two graphs: one for the pressure calculated from the ideal gas law and a second using the pressure from the VDW equation. Explain your data in terms of a transition from a gas to a supercritical fluid. Are the ideal gas law or the VDW valid above the T_c and P_c? Why or why not? Also, observe and explain how the CO_2 molecules fill the tube (*i.e.* align themselves) at the higher pressures. Include several images of the nanotube with different numbers of CO_2 molecules entrapped.

Finally, after each CO_2 is added and the parameters are calculated, measure several C=O bond distances and the O=C=O bond angles. Plot the pressure verses the average C=O bond distance. Is there a change in these parameters? Explain.

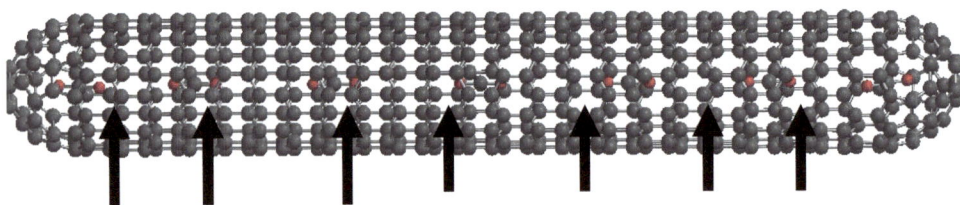

Figure 1: Seven CO_2 molecules are trapped in a carbon nanotube. Measure the distance between each adjacent molecule (1,2; 2,3; 3,4; 4,5; 5,6; 6,7) and average the values. Also calculate the pressure using the supercritical temperature.

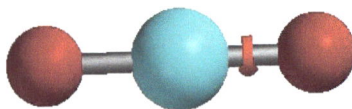

Figure 2: A Single CO_2 molecule. Measure the C=O bond distances and O=C=O bond angles after each calculation.

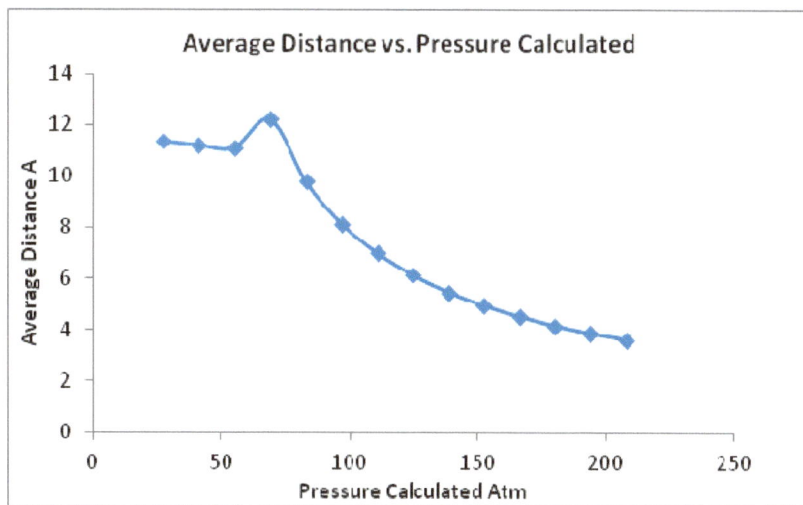

Figure 3: The average distance between CO_2 molecules trapped in a nanotube and the pressure cacualted with the Ideal Gas Law. There is a break at approximately the P_c.

REFERENCES

[1]   Tsai, P. J.; Yang, C. H.; Hsu, W. C.; Tsai, W. T.; Chang, J. K. Enhancing hydrogen storage on carbon nanotubes *via* hybrid chemical etching and Pt decoration employing supercritical carbon dioxide fluid, *Int. J. Hydrogen Energ.*, **2012**, *37*, 6714-6720.

[2]   Mandel, F. S.; Wang, J. D. Manufacturing of specialty materials in supercritical fluid carbon dioxide*, Inorg. Chim. Acta*, **1999**, *294*, 214-223.

[3]   Tello, J.; Viguera, M.; Calvo, L. Extraction of caffeine from Robusta coffee (Coffea canephora var. Robusta) husks using supercritical carbon dioxide, *J. Supercrit. Fluids*, **2011**, *59*, 53-60.

[4]   Kim, W. J.; Kim, J. D.; Kim, J.; Oh, S. G.; Lee, Y. W. Selective caffeine removal from green tea using supercritical carbon dioxide extraction, *J. Food Eng.*, **2008**, *89*, 303-309.

Send Orders of Reprints at reprints@benthamscience.net

CHAPTER 20

Radioactive Equilibrium

Thomas J. Manning* and Aurora P. Gramatges

Department of Chemistry, Valdosta State University, Valdosta, Georgia, USA, and Instituto Superior de Tecnología y Ciencias Aplicadas, La Habana, Cuba

Abstract: In this exercise students will learn the different types of radioactive equilibrium. The students will build graphs of radioactive decay using Excel function of multiple series, and will be able to compare the behavior of these processes according to the equilibrium type.

Keywords: Chemical kinetics, radioactive equilibrium, radioactive decay, nuclear transformations.

INTRODUCTION

The radioactive decay reaction is a typical example of a reaction with first order kinetics. In this exercise the students will learn about the different transformations that can occur in a mixture of radioisotopes that are genetically related in a decay reaction [1, 2]. This reaction has the same behavior than the classical consecutive reactions in chemical kinetics. The student will be familiar with some basic terminology used in nuclear chemistry, and will build graphs in Excel with the results obtained when the condition of radioactive equilibrium is reached. In this exercise, the student will also use the logarithmic scale in a multiple series graph.

The radioactive decay processes are the best known nuclear reactions, in which an unstable nucleus is transformed into a different element by emitting one or more particles [3]. In these processes the reactive is called *parent* nucleus and the product is called *daughter* nucleus. For example, in the nuclear decay equation of a uranium-235 nucleus, the parent uranium-235 is disintegrated to form the daughter nucleus thorium-231, by emitting an alpha particle [4]:

$$^{235}_{92}U \rightarrow {}^{231}_{90}Th + {}^{4}_{2}He^{2+}$$

*Address correspondence to Thomas J. Manning: Department of Chemistry, Valdosta State University Valdosta GA 31698, USA; Tel: 229-333-7178; E-mail: tmanning@valdosta.edu

The reaction rate of decay of a radioisotope is proportional to the number of atoms of that isotope present at that instant. Radioactive decay, thus, follows the first order kinetics: $-\dfrac{dN}{dt} = N\lambda$, where N is the number of atoms at any time t, and λ is the disintegration constant which is related to time by the equation: $\lambda = \dfrac{\ln 2}{T_{1/2}}$. (Note that in this case λ is used instead of k as the rate constant, and therefore it has dimensions of time^{-1}, *e.g.* s^{-1}.) The negative sign implies the decay of atoms. The product $N\lambda$ is known as radioactivity or just *activity* (*A*), and its value decreases exponentially as a function of time: $A = A_0 e^{-\lambda t}$. Activity has the unit of disintegrations per unit time, generally disintegrations per second (dps). The units of radioactivity are: 1 Becquerel (Bq) = 1 dps, 1 Curie (Ci) = 3.7 x 1010 dps. Similarly to first-order reactions, the time required for the decay of half of the parent atoms to daughter products is defined as the half-life (t1/2) of the parent nuclide.

In the radioactive decay processes there are two possibilities: either the product is stable or the product is radioactive. In the latter case, if the $T_{1/2}$ of the daughter is smaller than the $T_{1/2}$ of the parent, then the daughter activity grows with time according to the following equation:

$$A_2 = A_1^0 \left[\frac{\lambda_2}{\lambda_2 - \lambda_1} \right] \left[e^{-\lambda_1 t} - e^{-\lambda_2 t} \right]$$

where A_1^0 is the parent initial activity, and λ_1 λ_2 are the parent and daughter decay constants, respectively.

The total activity considering the parent disintegration and the daughter growing can be expressed as the sum of the activities of the parent and daughter nuclei:

$$A = A_1^0 e^{-\lambda_1 t} + A_1^0 \left[\frac{\lambda_2}{\lambda_2 - \lambda_1} \right] \left[e^{-\lambda_1 t} - e^{-\lambda_2 t} \right]$$

Depending on the disintegration constant of parent and daughter nuclei there are three typical cases of correlated disintegrations. The *secular equilibrium*, which occurs in those cases where the parent's t$_{1/2}$ is much higher than the daughter's, thus the disintegration constants follow λ_1 (parent) $\ll \lambda_2$ (daughter). The *transient equilibrium*, which occurs in those cases where the disintegration constants λ_1 (parent) and λ_2 (daughter) are in a 0.1 ratio. These two cases are known as radioactive equilibrium. In the cases where the daughter's $t_{1/2}$ is higher than the parent's, the equilibrium condition is not reached, and the daughter's activity increases up to a maximum and decays with its characteristic $t_{1/2}$.

Pre-Lab Questions

1. Which is the definition of half-life time of a radioactive decay reaction ($t_{1/2}$)? How is it related to the rate constant of this process?

2. Cesium-137 disintegrates to Barium-137m by emitting a beta particle. Considering that the $T_{1/2}$ of the reaction is 30.04 years, calculate the constant of radioactive disintegration (λ_1).

3. Cerium-144 disintegrates to Praseodinium-144, with a $t_{1/2}$ = 284.89 days, and this nucleus disintegrates in turn to Neodinium-144, with a $t_{1/2}$ = 17.28 min. Both reactions occur with emission of a beta particle. Is the radioactive equilibrium condition valid for these consecutive reactions? If that is the case, which type of equilibrium is shown in this reaction?

Laboratory Exercises

The name/title and pre-lab questions should take two pages (maximum) and your graph should be pasted into its own page with a figure caption (*i.e.* Fig. **1**. A graph of Activity *vs.* Time for the radioactive decay ....). After completing this exercise, you should work on the equilibrium in Part II, which is the other case of radioactive equilibrium. Note that although the procedure is similar, the graphs obtained are not the same and you should first obtain the general equation that is used for total activity as a function of time in the new equilibrium.

Remember to save your work (Excel files, report) to at least two memory devices (*i.e.* memory stick, hard disk) on a regular basis! Many computers at universities and libraries have programs installed that will delete your file automatically for security reasons.

Part I

A typical example of secular equilibrium is the disintegration reaction: ^{137}Cs ➔ ^{137m}Ba. $t_{1/2}(^{137}Cs) = 30.04$ years and $t_{1/2}(^{137m}Ba) = 2.55$ min. Consider that the initial activity of ^{137}Cs is $A_1^0(^{137}Cs) = 1$ mCi.

1. Open a new spreadsheet in Excel. In the box A1 enter the title "Time". In box A2 enter the number zero "0". In box A3 enter the command "= SUM(A2+0.1)" and copy/paste down to the box A302. Verify that in A302 appears the number "200". These values represent the time intervals at which the reaction rate is measured.

2. In box B1 enter "Parent Disint. Constant". In box B2 enter "=LN(2)/(30.04*365*24*60)". This is the value for the rate constant of the reaction ^{137}Cs ➔ ^{137m}Ba, after converting 30.04 years into minutes. Copy this value and paste it down to box B302.

3. In box C1 enter "Daughter Disint. Constant". In box C2 enter "=LN(2)/2.55". This is the value for the rate constant of the reaction ^{137}Cs ➔ ^{137m}Ba. Copy this value and paste it down to box C302.

4. In box D1 enter the title "Parent Act". and in box D2 enter the numeric value "2.22E+09" (A_0, initial activity of the parent-^{137}Cs). In box D3 enter the command "=$D$2*EXP(-B3*A3)" and copy/paste down to box D302. Note that the numerical values for the parent's activity are the same, since the half-life of the parent is long enough so the activity remains practically unchanged during the time interval we are using (30 min). **Question:** What would be the parent's activity after 20 years?

5. In box E1 enter the heading "Daughter Act". and in box E2 enter the value zero "0". This is the initial activity of ^{137m}Ba (Daughter). In box E3 enter the command "=$D$2*(C3/(C3–B3))*(EXP(-B3*A3)-EXP(-C3*A3))" and copy/paste down to box E302.

6. In box F1 enter the title "Parent + Daughter Act". And in box F2 enter the numeric value "2.22E+09". Since we start with a pure sample, at time zero the total activity is equal to the initial activity (A_0) of ^{137}Cs (parent), expressed in Becquerel.

7. Now we are going to calculate the total activity (Parent + Daughter) at different times, using Eq. 5.2. In the box F3 enter the command "=$F$2*EXP(-B3*A3) +$F$2*(C3 /(C3 −B3))*(EXP(-B3*A3)-EXP(-C3*A3))" and copy/paste down to the box F302.

8. Let's now plot the data of these three curves representing the radioactive equilibrium. This exercise assumes that you are familiar with creating graphs in Excel. Name the first series as "Parent Act"., the second series as "Daughter Act"., and the third series as "Parent + Daughter Act".. Use the values in D2…D302, E3…E302 and F2…F302 for the y-axis in each series, respectively, and the values in A2…A302 for x-axis in the three series. Once the graph is completed, adjust the scale in the y-axis in the range 1E+09 to 1E+10 and **select the logarithm option** for this axis**.** Adjust the x-axis scale in the range 0 to 30. (Note that in the case of the values for Daughter Activity, the initial value is not used in the graph, since logarithm of zero is not defined.) Enter a title for the graph ("Secular Equilibrium"), as well as for the x-axis ("Time (h)") and for the y-axis ("Activity (in log scale)"). Deselect any background color of the graph. Select the secondary gridlines in the graph menu. Copy and paste your graph in the report and write an appropriate figure caption. Your graph should look like the one in Fig. **1**.

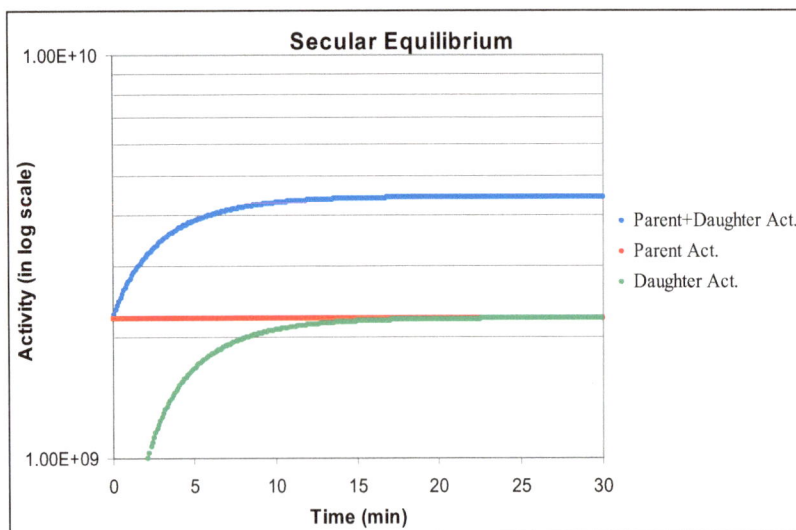

Figure 1: Graph of Activity *vs.* Time of the decay reaction ^{137}Cs ➔ ^{137m}Ba.

Part II

Create in Excel the radioactive decay curves that represent the equilibrium ^{99}Mo → ^{99m}Tc. ($T_{1/2}(^{99}$Mo) = 66.7 h and $t_{1/2}(^{99m}$Tc) = 6.01 h). Consider the initial activity of ^{99}Mo as $A_1^0(^{99}$Mo) = 1 mCi. (Remember that 1 Ci = 1.37×10^{10} Bq = 1.37×10^{10} disintegrations/second).

Hint: In the case of transient equilibrium, the total activity (sum of parent's plus daughter's activities in a sample that initially only contains the parent nuclide) goes through a maximum before the equilibrium condition is reached.

Post-Lab Questions

1. Why is it necessary to express the activity in Becquerel in the equation of radioactive decay?

2. Regular chemical reaction rate constants depend on temperature and pressure, among other conditions. Do you think that radioactive decay rate constants are also modified by changes in external conditions? Explain.

3. Given the reaction of radioactive decay: ^{122}Xe → ^{122}I → ^{122}Te (stable). Obtain the activity that is present in an initially pure sample containing 1 Ci of ^{122}Xe, after 2 min have elapsed. $t_{1/2}(^{122}$Xe) = 20.1 h, $t_{1/2}(^{122}$I) = 3.6 min.

REFERENCES

[1]   Breger, I. A. Radioactive equilibrium in ancient marine sediments, *Geochim. Cosmochim. Acta*, **1955**, *8*, 63-73.
[2]   Santos, R..; Marques, L. S. Investigation of ^{238}U–^{230}Th–^{226}Ra and ^{232}Th–^{228}Ra–^{228}Th radioactive disequilibria in volcanic rocks from Trindade and Martin Vaz Islands (Brazil; Southern Atlantic Ocean), *J. Volcanol. Geoth. Res.*, **2007**, *3*, 215-233.
[3]   Kónya, J.; Nagy, N. M. *Nuclear and Radiochemistry*, Elsevier, Oxford, **2012**.
[4]   Kalmaz, E. V.; Barbieri, J. L. Mathematical modelling and computer simulation of radioactive and toxic chemical species dispersion in porous media in the vicinity of uranium recovery operations, *Ecol. Model.*, **1981**,*13*, 159-181.

Send Orders of Reprints at reprints@benthamscience.net

CHAPTER 21

Geography and the Global Chemical Market

Thomas J. Manning* and Aurora P. Gramatges

Department of Chemistry, Valdosta State University, Valdosta, Georgia, USA, and Instituto Superior de Tecnología y Ciencias Aplicadas, La Habana, Cuba

Abstract: This is an interdisciplinary exercise that encompasses geography, chemistry, health care and touches on a number of other topics (geology, agriculture, *etc.*). Students will examine the location of a country and its natural resources that can be related to some facet of the chemical industry. Students correlate the development of the chemical industry with life expectancy.

Keywords: World chemical market, geography, chemical industry, correlations.

INTRODUCTION

In most chemistry class's topics such as chemical bonding, spectroscopy, kinetics, synthesis's, stoichiometry and thermodynamics dominate lectures, labs and homework assignments. This interdisciplinary exercise introduces the impact that the chemical industry has on the global economy [1-4]. Students are provided with a list of countries, asked to locate them on a map, identify three neighboring countries and identify 1-3 products or natural resources produced by that country. The chemical industry touches all of the major industries including petrochemical, agriculture, mining, transportation, pharmaceutical and specialty products. While a country may not seem technologically advanced, it may contain a natural resource (ores, timber, livestock, *etc.*) that are part of the global chemical industry. An area such as agriculture shows how many different areas of chemistry are involved in a major market including genetics, production of herbicides and pesticides, production of fertilizers and the quality control testing of the food.

In this exercise students will complete Table **1**. Recommended web sites include

---

*Address correspondence to Thomas J. Manning:** Department of Chemistry, Valdosta State University Valdosta GA 31698, USA; Tel: 229-333-7178; E-mail: tmanning@valdosta.edu

1. The CIA Fact book at https://www.cia.gov/library/publications/the-world-factbook/

2. Wikipedia list of countries at: http://en.wikipedia.org/wiki/List_of_countries

There are many resources on the web related to this exercise but the two sites above are easy to navigate. In filling in Table **1**, identify the three neighboring countries. For an island country, simply select the three closest countries (3rd column). In listing the natural resources (4th column), pick three products that are exported and may be involved in the chemical industry. Smaller countries may only export some foodstuffs but these may be grown, harvested and stored efficiently by herbicides and pesticides or be preserved by chemicals. Larger countries may have well developed commercial enterprises in many areas but simply pick three larger industries. In the fifth and final column record the average life expectancy of that country. This number provides an insight into how advanced countries basic resources are, including the quality of water it provides for its citizens, the nutritional level available, and the health of infant and small children. When you consider that a woman in Japan today will live to be an average of 82 years old while a women born in Swaziland (located in southern Africa) will live to be less than half that age (39 years old), you begin to understand the positive impact that the chemical industry can have on the quality of life.

Your instructor may choose to reduce the total number of countries you examine (*i.e.* just odd numbered countries or every 3rd country on list, *etc.*) but when this is complete you should have gained a better insight to the world in which we live and the impact that chemistry has on your global society.

Table 1: Fill in this table by hand using the web sites indicated above (or others).

#	Country	Three Neighboring Countries	Chemically Related Products	Life Span
1	Afghanistan			
2	Algeria			
3	Angola			

Table 1: contd....

4	Argentina			
5	Armenia			
6	Austria			
7	Australia			
8	Bahamas			
9	Bahrain			
10	Bangladesh			
11	Barbados			
12	Belgium			
13	Belize			
14	Bermuda			
15	Bhutan			
16	Botswana			
17	Brazil			
18	British Virgin Islands			
19	Brunei			
20	Burkina Faso			
21	Burundi			
22	Cambodia			
23	Cameroon			
24	Canada			
25	Central African Republic			
26	Chile			
27	China (People's Republic)			
28	Colombia			
29	Congo (Republic of the Congo)			
30	Costa Rica			
31	Croatia			
32	Cuba			
33	Cyprus			
34	Czech Republic			
35	Denmark			
36	Djibouti			

Table 1: contd….

37	Dominican Republic			
38	Ecuador			
39	Egypt			
40	Equatorial Guinea			
41	Estonia			
42	Fiji			
43	Finland			
44	France			
45	Gabon			
46	Greece			
47	Guatemala			
48	Guinea			
49	Guinea-Bissau			
50	Guyana			
51	Haiti			
52	Honduras			
53	Hong Kong			
54	Hungary			
55	Iceland			
56	India			
57	Indonesia			
58	Iran			
59	Iraq			
60	Ireland			
61	Israel			
62	Italy			
63	Jamaica			
64	Japan			
65	Kazakhstan			
66	Kenya			
67	South Korea			
68	Kuwait			
69	Kyrgyzstan			

Table 1: contd….

70	Lebanon			
71	Libya			
72	Macedonia			
73	Madagascar			
74	Malaysia			
75	Mali			
76	Mexico			
77	Micronesia			
78	Mongolia			
79	New Zealand			
80	Nicaragua			
81	Niger			
82	Nigeria			
83	North Korea			
84	Norway			
85	Pakistan			
86	Panama			
87	Papua New Guinea			
88	Peru			
89	Philippines			
90	Qatar			
91	Romania			
92	Russia			
93	Rwanda			
94	Samoa			
95	Saudi Arabia			
96	Senegal			
97	Serbia and Montenegro			
98	Sierra Leone			
99	Singapore			
100	Slovenia			
101	Solomon Islands			
102	Somalia			

Table 1: contd….

103	South Africa			
104	Spain			
105	Sri Lanka			
106	Sudan			
107	Swaziland			
108	Sweden*			
109	Syria			
110	Taiwan			
111	Timor-Leste (East Timor)			
112	Togo			
113	Tonga			
114	Turkey			
115	Turkmenistan			
116	Turks and Caicos Islands			
117	Uganda			
118	Ukraine			
119	United Arab Emirates			
120	Great Britain			
121	United States of America			
122	Uruguay			
123	Uzbekistan			
124	Venezuela			
125	Viet Nam			
126	Yemen			
127	Zimbabwe			

REFERENCES

[1]   Lotze-Campen, H.; Popp, A.; Beringer, T.; Müller, C.; Bondeau, A.; Rost, S.; Lucht, W. Scenarios of global bioenergy production: The trade-offs between agricultural expansion, intensification and trade, *Ecol. Model.*, **2010**, *221*, 2188-2196.

[2]   Vertova, G. A historical investigation of the geography of innovative activities, *Struct. Change Econ. Dynamics*, **2002**, *13*, 259-283.

[3]   Cooke, P. Regionally asymmetric knowledge capabilities and open innovation: Exploring 'Globalization 2'—a new model of industry organization, *Res. Policy*, **2005**, *34*, 1128-1149.

[4]   Achilladelis, B.; Antonakis, N. The dynamics of technological innovation: the case of the pharmaceutical industry, *Res. Policy*, **2001**, *30*, 535-588.

CHAPTER 22

A Periodic Puzzle

Thomas J. Manning* and Aurora P. Gramatges

Department of Chemistry, Valdosta State University, Valdosta, Georgia, USA, and Instituto Superior de Tecnología y Ciencias Aplicadas, La Habana, Cuba

Abstract: This is a Sudoku type exercise that focuses on periodic trends and elemental symbols. It does require access to an interactive periodic table. It is difficult!

Keywords: Periodic properties, periodic trends, chemical symbols, Sudoku, logical calculations.

INTRODUCTION

The logic behind Latin squares (Fig. **1**) has been popularized by the Sudoku puzzle series. Sudoku appears in many newspapers, on-line forums and books of puzzles can be purchased for the popular numerical game [1-3]. This exercise shares some general similarities but is different in a number of aspects including the fact that it can only be solved with 81 different elemental symbols, an allocation of spaces is done by periodic groups and alphabetical considerations.

Recognizing patterns is fundamental to many areas of chemistry and molecular structures. Natural polymers such as DNA, RNA, proteins, cellulose and lignin all have recognizable patterns. Salts and materials such as NaCl, CsCl and graphite have repeating units that define its structure. To solve this puzzle, the best approach is to develop and apply a series of nine patterns (nine rows). Along the way, students will be forced to examine the periodic table closely and to recognize different groups [4].

Fig. **1** provides a simple example of a simple 3x3 Latin square with 3 symbols (1,2,3).

1	2	3
3	1	2
2	3	1

*Address correspondence to **Thomas J. Manning:** Department of Chemistry, Valdosta State University Valdosta GA 31698, USA; Tel: 229-333-7178; E-mail: tmanning@valdosta.edu

Students are directed to a web site or are given a quick lecture about the rules of Sudoku. Once the logic of the popular puzzle is explained, the groups (see Table 2) and the rules used in this exercise are outlined.

Table 1: The nine groups used periodic puzzles.

A.	The element is a gas at 1 atm and 0°C.
B.	The elements have a stable oxidation state of +1 in a complex, salt or dissolved in water.
C.	The element is one of the lanthanides (La-Lu).
D.	The element is one of the actinides (Ac-Lr).
E.	The element has a stable oxidation state of +2 (complex, salt, dissolved in water).
F.	The element is a nonmetal or a metalloid (all are to the right of the metalloid break).
G.	The element is a soft metal or metalloid (left of the metalloid break) or a transition metal with a 4d outer orbital (Y-Cd).
H.	The element is a transition metal with an outer 5d orbital (Hf-Hg).
I.	The element is an artificial element with 104 to 118 protons (Rf-Uuo).

A. Full Periodic Puzzle

The rules for the Periodic Puzzle format are:

1. Each 3x3 block contains an element from each of the nine groups listed above. Each correctly solved grid will have 81 different elemental symbols (3 x 3 x 9).

2. There cannot be two elements from the same group (A-I above) in the same row or column (vertical, horizontal). An element may qualify in two groups (Cl as nonmetal or as a gas) but once it is used in a specific group, it does not apply to the second group.

3. Write in the element symbol only (no charge, state, subscripts, *etc.*).

4. http://www.dayah.com/periodic/ This periodic table lists all of the elements that can be used in this grid. If the symbol is clicked on, it provides links to stable oxidation states.

5. Symbols for species with up to 118 protons are possible.

6. Multiple symbols may be possible for a specific box (choose wisely!).

7. Each element can be used only once in the entire 9x9 puzzle. The final puzzle should have 81 different symbols.

8. Some elements have the potential to be in different groups (*i.e.* Cl can be a nonmetal or a gas). Once you use an element in one group it cannot be used in another group.

9. Hydrogen (H), Deuterium (D), and Tritium (T) are isotopes but are treated as separate species for potential use as a gas (T_2, D_2, H_2), a singly charged ion (T^+, D^+, H^+) or a nonmetal (T, D, H).

10. For a group designated by an oxidation number, it is a stable species when dissolved in water, part of a salt or part of a complex. It does not have to be a species with only one stable oxidation state (Fe^{+2}, Fe^{+3}).

11. Each row and column can only have one element with a first letter (*i.e.* sulfur (S) and samarium (Sm) cannot be in the same row or column). Each symbol in any vertical or horizontal list must start with nine different letters.

12. There are no diagonal constraints with letters or groups.

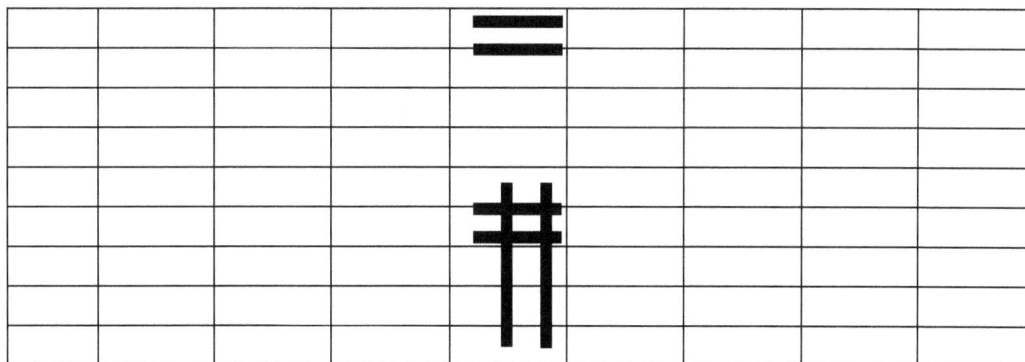

Figure 1: This empty 9x9 grid contains nine 3x3 sub-grids. Along with the 9x9 grid rules, access to the on-line periodic table, and the nine elemental groups (Table 1), this grid is provided to participants.

Typically, the best approach to solve this puzzle is to develop a logic pattern based on groups. For example, the first row across would be an element from groups A,B,C,D,E,F,G,H,I in consecutive boxes and the second row across would be elements from G,H,I,A,B,C,D,E,F in consecutive boxes, *etc.* Once this is accomplished they can be rearranged to account for no letters being the same [6]. If needed a solved copy of this puzzle can be obtained by e-mailing tmanning@valdosta.edu.

REFERENCES

[1]   Crook, J. F. A Pencil-and-Paper Algorithm for Solving Sudoku Puzzles, *Not. Am. Math. Soc.,* **2009**, *56*, 460-468.

[2]   Mepham, M. (), *Solving Sudoku*, Crosswords Ltd., Frome, England, **2005**. This article is available online at http://www.sudoku.org.uk/PDF/Solving_Sudoku.pdf.

[3]   Sheldon, T. *Sudoku Master Class*, Plume, Penguin, New York, **2006.**

[4]   Chemicool Periodic Table. http://www.chemicool.com/ (accessed on June 16, **2012**).

[5]   Manning, T, Periodic Puzzles: The Periodic Table, Element Symbols and Pattern Recognition, Chem. Ed 14, 1-3 **2009**. (see for additional details)

[6]   http://www.valdosta.edu/periodicpuzzles/.

CHAPTER 23

Count the Sodium Atoms

Thomas J. Manning[*] and Aurora P. Gramatges

Department of Chemistry, Valdosta State University, Valdosta, Georgia, USA, and Instituto Superior de Tecnología y Ciencias Aplicadas, La Habana, Cuba

Abstract: This exercise aims to teach students chemical nomenclature and empirical formulas, as well as some concepts of isomers and polarity.

Keywords: Sodium, minerals, coordination compounds, nomenclature, empirical formula.

INTRODUCTION

I. Sodium Compound Nomenclature and Empirical Formulas

This exercise is focused on nomenclature and empirical formulas. Student may use any resource (*i.e.* web, textbooks, instructor) to find the empirical formulas. In column A is a list of 100 simple salts, coordination compounds and minerals. Students find the empirical formula and place it in column B. In column C they write out the number of sodium atoms. All compounds listed have at least one sodium ion in their empirical formula, while some may have several. At the bottom of Table **1**, put in your total number of sodium ions for all compounds. In order to get full credit, it is important to find all of the empirical formulas. This exercise may be used in a chemistry or geology course and maybe used to introduce some aspects of nomenclature or to refresh a student on the topic.

For the minerals it is important to note that you must find the empirical formula. For example, FERRO-EDENITE (Sodium Calcium Iron Magnesium Aluminum Silicate Hydroxide), the elemental names are likely NOT 1:1:1, *etc.* The name just gives you the elements present.

The table can only be completed by hand (pen, pencil). Typed sheets will not be accepted.

---
*Address correspondence to Thomas J. Manning:** Department of Chemistry, Valdosta State University Valdosta GA 31698, USA; Tel: 229-333-7178; E-mail: tmanning@valdosta.edu

Table 1: Empirical formula and number of sodium atoms of simple salts, coordination compounds and minerals.

#	Name	Empirical Formula	# Sodium Atoms
1	Sodium Hydride		
2	Sodium Oxide		
3	Sodium Fluoride		
4	Sodium Sulfide		
5	Sodium Nitride		
6	Sodium Bromide		
7	Sodium Iodide		
8	Sodium Chloride		
9	Sodium Nitrate		
10	Sodium Phosphate		
11	Sodium Sulfate		
12	Sodium Perchlorate		
13	Sodium Iodate		
14	Sodium Chlorate		
15	Sodium Bromate		
16	Sodium Carbonate		
17	Sodium Bicarbonate		
18	Sodium Arsenate		
19	Sodium Arsenite		
20	Sodium Hydrogen Phosphate		
21	Sodium Dihydrogen Phosphate		
22	Sodium Hydrogen Sulfate		
23	Sodium Nitrite		
24	Sodium Thiosulfate		
25	Sodium Sulfite		
26	Sodium Chlorite		
27	Sodium Hypochlorite		
28	Sodium Hypobromite		
29	Sodium Chromate		
30	Sodium Dichromate		
31	Sodium Acetate		
32	Sodium Formate		

Table 1: contd...

33	Sodium Cyanide		
34	Sodium Amide		
35	Sodium Cyanate		
36	Sodium Peroxide		
37	Sodium Thiocyanate		
38	Sodium Oxalate		
39	Sodium Hydroxide		
40	Sodium Permanganate		
41	Sodium Amminetrichloroplatinate(II)		
42	Sodium Tetrabromocuprate(II)		
43	Sodium Pentachloronitridoosmate(VI)		
44	Sodium Tetrachloronickelate(II)		
45	Sodium Diaquabis(oxalato)nickelate(II)		
46	Sodium Tetrachlorocuprate(II)		
47	Sodium Monochloropentacyanoferrate(III)		
48	Sodium Hexafluorocobaltate(III)		
49	AEGIRINE (Sodium Iron Silicate)		
50	ALBITE (Sodium Aluminum Silicate)		
51	BUERGERITE (Sodium Iron Aluminum Boro-silicate Oxide Fluoride)		
52	BURBANKITE (Sodium Calcium Strontium Barium Cerium Carbonate)		
53	BYTOWNITE (Calcium Sodium Aluminum Silicate)		
54	CANCRINITE (Sodium Calcium Aluminum Silicate Carbonate)		
55	CARBOCERNAITE (Calcium Sodium Strontium Cerium Barium Carbonate)		
56	CATAPLEIITE (Hydrated Sodium Zirconium Silicate)		
57	CHAROITE (Hydrated Sodium Calcium Barium Strontium Silicate Hydroxide Fluoride)		
58	CHKALOVITE (Sodium Beryllium Silicate)		
59	CLINOPTILOLITE (Hydrated Sodium Potassium Calcium Aluminum Silicate)		
60	CROCIDOLITE (asbestos form variety of Riebeckite, Sodium Iron Magnesium Silicate Hydroxide)		
61	CRYOLITE (Sodium Aluminum Fluoride)		
62	DRAVITE (complex Sodium Magnesium Iron Boro-Aluminum Silicate)		

Table 1: contd...

63	EDENITE (Sodium Calcium Magnesium Iron Aluminum Silicate Hydroxide)		
64	ELBAITE (Sodium Lithium Aluminum Boro-Silicate Hydroxide)		
65	ELPIDITE (Hydrated Sodium Zirconium Silicate)		
66	EPIDIYMITE (Sodium Beryllium Silicate Hydroxide)		
67	ERIONITE (Hydrated Potassium Sodium Calcium Magnesium Aluminum Silicate)		
68	EUDIALYTE (Sodium Calcium Cerium Iron Manganese Zirconium Silicate Hydroxide Chloride)		
69	EUDIDYMITE (Sodium Beryllium Silicate Hydroxide)		
70	FEDORITE (Hydrated Sodium Potassium Calcium Silicate Fluoride Chloride Hydroxide)		
71	FERRO-EDENITE (Sodium Calcium Iron Magnesium Aluminum Silicate Hydroxide)		
72	FERROGLAUCOPHANE (Sodium Iron Magnesium Silicate Hydroxide)		
73	FLUORRICHTERITE (Sodium Calcium Magnesium Iron Silicate Hydroxide Fluoride)		
74	GAYLUSSITE (Hydrated Sodium Calcium Carbonate)		
75	GLAUBERITE (Sodium Calcium Sulfate)		
76	GLAUCOPHANE (Sodium Magnesium Iron Aluminum Silicate Hydroxide)		
77	GMELINITE (Hydrated Sodium Calcium Aluminum Silicate)		
78	HACKMANITE (variety of Sodalite, Sodium Aluminum Silicate Chloride)		
79	HALITE (Sodium Chloride)		
80	HANKSITE (Potassium Sodium Sulfate Carbonate Chloride)		
81	HEULANDITE (Hydrated Sodium Calcium Aluminum Silicate)		
82	HILAIRITE (Hydrated Sodium Zirconium Silicate)		
83	JADEITE (Sodium Iron Aluminum Silicate)		
84	JOAQUINITE (Barium Sodium Cesium Titanium Niobium Iron Silicate Hydroxide Fluoride)		
85	KERNITE (Hydrated Sodium Borate)		
86	KIDWELLITE (Hydrated Sodium Iron Phosphate Hydroxide)		
87	KUPLETSKITE (Potassium Sodium Manganese Iron Titanium Niobium Silicate Hydroxide)		
88	LAZURITE (Sodium Calcium Aluminum Silicate Sulfate)		

89	LEIFITE (Hydrated Sodium Beryllium Aluminum Silicate Hydroxide Fluoride)		
90	LORENZENITE (Sodium Titanium Silicate)		
91	MESOLITE (Hydrated Sodium Calcium Aluminum Silicate)		
92	MICROLITE (Calcium Sodium Tantalum Oxide Hydroxide Fluoride)		
93	MONTEBRASITE (Lithium Sodium Aluminum Phosphate Hydroxide Fluoride)		
94	MONTMORILLONITE (Hydrated Sodium Calcium Aluminum Magnesium Silicate Hydroxide)		
95	NAHCOLITE (Sodium Bicarbonate)		
96	NARSARSUKITE (Sodium Titanium Iron Silicate Fluoride)		
97	NATROJAROSITE (Sodium Iron Sulfate Hydroxide)		
98	NITRATINE (Sodium Nitrate)		
99	OLIGOCLASE (Sodium Calcium Aluminum Silicate)		
100	PECTOLITE (Sodium Calcium Silicate Hydroxide)		

Add your total number of sodium atoms and record here.

Total # of Sodium atoms: _____

II. Sodium Compound Isomers and Polarity

Students will know choose ten of the sodium compounds from Table **1**, preference is given to isomers. The compounds will be built in Spartan and the dipole moment will be calculated for each the results can be inserted into Table **2**.

Table 2: Spartan image and dipole moment of select sodium compounds.

Sodium Compound	Spartan Image	Dipole Moment

Send Orders of Reprints at reprints@benthamscience.net

CHAPTER 24

Computational Work with Natural Products

Thomas J. Manning* and Aurora P. Gramatges

Department of Chemistry, Valdosta State University, Valdosta, Georgia, USA, and Instituto Superior de Tecnología y Ciencias Aplicadas, La Habana, Cuba

Abstract: This exercises aims to familiarize students with chemical structures of natural products. Students will also use computational methods to learn about trends or the lack thereof in natural products.

Keywords: Natural product, computational calculation, chemical properties.

INTRODUCTION

A natural product is a molecule derived from natural sources. Natural products are extensively studied for their pharmaceutical activity. Approximately forty seven percent of small molecule anticancer drugs are or are derived from natural products [1]. Natural products consist of five major structural groups: polyphenols, natural phenols, phytosterols, terpenoids and alkaloids. Examples of alkaloids include the painkiller morphine, the anticancer drug vincristine and the antimalarial drug quinine. An example of a natural phenol is the molecule responsible for the health benefits of red wine, resveratrol [2].

The bestselling cancer drug of all times, Paclitaxel, is a natural product that was originally isolated from the bark of the pacific yew tree. Its anticancer properties were identified in 1971. Paclitaxel is not isolated from the pacific yew today, but discovery of its anticancer properties led to investment in a method of synthesis for this important pharmaceutical agent. Bristol-Myers-Squibb commercially developed Paclitaxel and it became available to the public in 1992 [3].

Acetylsalicylic acid, also known as aspirin, has analgesic (pain relieving), antipyretic (fever reducing) and anti-inflammatory properties. Aspirin was

---

*Address correspondence to Thomas J. Manning:** Department of Chemistry, Valdosta State University Valdosta GA 31698, USA; Tel: 229-333-7178; E-mail: tmanning@valdosta.edu

isolated from willow bark in the early nineteenth century, but Hippocrates was known to use a powder from the willow tree for reducing pain and fever. Hippocrates was alive as early as 460 B.C. Although the properties of willow bark have been known for thousands of years, aspirin was first commercialized by Bayer in 1900 [4].

Natural products are continuously being discovered and serve as strong inspiration for modern pharmaceutical development.

II. The Project

The following exercise is focused on using computational methods to learn about trends or the lack thereof in natural products. In addition to researching some background information on the drug, students will build it in Spartan. This would be an exhaustive exercise for a single student to do but would make an excellent group or class project. If completed as a group project and most or all of the molecules are finished, than students can use this data to propose a single "average" molecule.

For the average molecule, students pick a disease that is well represented in the list of medications (*i.e.* cancer) and use the data from these drugs. First they will determine the average empirical formula by using the values from each element (C, H, N, O, *etc.*). This value is a starting point for building their structure. Next they determine the average number of each functional group (*i.e.* amine, carbonyls,). Typically they will round up this number. For example, if the average natural product used to treat cancer in their list has 0.6 esters, they build a molecule and include a single ester. Other parameters such as dipole moment, surface area, chiral centers, volume and rings are also considered. These numbers are all guidelines and not absolutes. Once this structure is complete, build it in Spartan and calculate your final parameters. There are many possibilities.

Table **1** lists approximately fifty well known natural products that have been used in medicine. Literally hundreds of thousands of natural products have been discovered but not all have the same level of pharmaceutical activity so only the best emerge. Once this exercise is done you may see some structural trends emerge that most drugs have and chemical characteristics none have. Also, since most drugs need some polarity to work in the human body, nonpolar natural products are typically less likely to be successful (but not always).

Table 1: Information about fifty natural products.

	Empirical Formula	Molar Mass	Natural Source	Disease(s) Treated	Hand Drawn Structure	Spartan Structure
Aspirin						
Alpidine						
Amrubicin						
Anidulafungin						
Apomorphine						
Artemotil						
Aztreonam						
Biapenem						
Bivalirudin						
Bryostatin						
Bleomycin						
Capsaicin						
Caspofungin						
Cefditoren						
Codinaeopsin						
Colchicine						
Daptomycin						
Dimethyltryptamine						
Doripenem						
Dronabinol						
Ergotamine						
Ertapenem						
Everolimus						
Exenatide						
Fumagillin						
Galantamine						
Ixabepilone						
Lisdexamfetamine						
Mescaline						
Methylnaltrexone						
Micafungin						

Miglustat					
Mycophenolate					
Nitisinone					
Orlistat					
Paclitaxel					
Phenethylamine					
Pimecrolimus					
Retapamulin					
Romidepsin					
Rosuvastatin					
Spiruchostatins					
Telavancin					
Telithromycin					
Temsirolimus					
Tigecycline					
Tiotropium					
Trabectedin					
Ziconotide					
Zotarolimus					
Quinine					
ET743					

Averaged molecule: Disease:

Average Number of Atoms Table

Atom	C	N	H	O	S	Halogen
Average						

Empirical Formula

Average Number of Functional Groups Table

Group Name							
Average							

Calculated Parameters Table

Disease	Hand Drawn Structure	Spartan Structure	Calculated Dipole Moment	Calculated Volume	Dipole Moment/Volume Ratio	Surface Area

REFERENCES

[1]   Newman, D.; Cragg, G., Natural products as sources of new drugs over the last 25 years, *J. Nat. Prod.*, **2007**, *70*(3), 461-477.

[2]   Lopez-Velez, M.; Martinez, F.; Del Valle-Ribes, C., The study of phenolic compounds as natural antioxidants in wine, *Critical Reviews in Food Science and Nutrition*, **2003**, *43*(2), 233-244.

[3]   Kingston, D., Taxol, A molecule for all seasons, *Chem. Commun.*, **2001**, *10*, 867-880.

[4]   Mueller, R.; Scheidt, S., History of drugs for thrombotic disease: Discovery, development, and directions for the future, *Circulation*, **1994**, *89*, 432-449.

Subject Index

www.ingramcontent.com/pod-product-compliance
Lightning Source LLC
Chambersburg PA
CBHW050830220326
41598CB00006B/339